Calculating Chance: Card and Casino Games

Sidney A. Morris

Calculating Chance: Card and Casino Games

Springer

Sidney A. Morris
Federation University
Ballarat, VIC, Australia

ISBN 978-3-031-70140-5   ISBN 978-3-031-70141-2 (eBook)
https://doi.org/10.1007/978-3-031-70141-2

Mathematics Subject Classification: 91A05, 33B15, 60-01

© The Editor(s) (if applicable) and The Author(s), under exclusive license to Springer Nature Switzerland AG 2024

This work is subject to copyright. All rights are solely and exclusively licensed by the Publisher, whether the whole or part of the material is concerned, specifically the rights of translation, reprinting, reuse of illustrations, recitation, broadcasting, reproduction on microfilms or in any other physical way, and transmission or information storage and retrieval, electronic adaptation, computer software, or by similar or dissimilar methodology now known or hereafter developed.
The use of general descriptive names, registered names, trademarks, service marks, etc. in this publication does not imply, even in the absence of a specific statement, that such names are exempt from the relevant protective laws and regulations and therefore free for general use.
The publisher, the authors and the editors are safe to assume that the advice and information in this book are believed to be true and accurate at the date of publication. Neither the publisher nor the authors or the editors give a warranty, expressed or implied, with respect to the material contained herein or for any errors or omissions that may have been made. The publisher remains neutral with regard to jurisdictional claims in published maps and institutional affiliations.

Cover illustration: Use Math_G

This Springer imprint is published by the registered company Springer Nature Switzerland AG
The registered company address is: Gewerbestrasse 11, 6330 Cham, Switzerland

If disposing of this product, please recycle the paper.

Dedicated to my wife

Elizabeth,

my daughter and son-in-law

Nasiya and Shimon,

and my grandchildren

Jonah, Georgia, and Lily

Foreword

I am delighted and privileged to write this foreword to Sid Morris's engaging book "Calculating chance: card and casino games". Sid and I were both students at the University of Queensland in the 1960s and have kept contact ever since through our common mathematical interests and links with the Australian Mathematical Society. Sid is an immensely successful author of scholarly mathematics texts, and now "Calculating chance" offers entertaining and informative reading for everyone. The book is immediately engaging. It is both personal and scholarly. At every stage Sid will give you the historical context, or the wider impact on society—and sometimes a very personal perspective. For example, did you know about the role that correct statistical thinking played in saving the lives of many airmen during World War II? In "Calculating chance" you will likely find every game of chance you've ever heard of in the book. You will discover its origins, you will meet the key people involved in its development, and Sid will explain how the game works and what are the chances of winning. And just when you think it is all for fun, you will be shown that the same mathematics explains how over 10,000 monogenic human diseases are passed down through families. Sid gives the clearest explanation I have ever read of the mathematics behind the Monty Hall Problem—a famous and confusing problem about a talk show program where you, the contestant, might win a car if out of three "doors" you choose the right door with a car behind it. After making your choice, the talk show host, who knows where the car is, opens a door that you did not choose, and the car is not behind it. So, either you chose correctly, or the car is behind the other door which the host did not open. The host offers you a chance to change your choice of door. What should you do? How does this opportunity change your chance of winning? Sid explains very clearly what is going on. Then, in the same part of the book you will find a fascinating mathematical analysis of the reliability of PCR tests for viral infection. It is the same kind of mathematics, and Sid's explanations lead on to discussions of different statistical approaches: Kolmogorov versus Bayes. So, you will find all you ever wanted, or needed, to know about games of chance – how the betting works, who benefits (always "the house"), and always there is the crucial role of mathematics to help you understand. It's fun to read, and there are problems to challenge you. The book is beautifully presented and illustrated: a "must-have" for your coffee table if you can bear to stop reading and put it down. Happy reading!

Emeritus Professor of Mathematics   Cheryl E Praeger AC FAA
University of Western Australia

Preface

This book is not written for any existing probability course as I wanted to decide which topics to include, and in what order, and in how much detail the material should be presented.

The following people should find this book interesting, entertaining, and/or useful:
- Students studying probability
- Those planning to study probability,
- Teachers of probability,
- Those interested in card games, casino games, or any games of chance

I assume a minimal background knowledge in mathematics and probability. The presentation is both gentle and rigorous. So I take the time to introduce and discuss necessary background material.

Many books separate probability into discrete distributions and continuous distributions. In so doing, they often gloss over an understanding of infinite set theory. I prefer dividing probability into two sections, not two, as I subdivide discrete distributions into finite and countably infinite. In this book we focus on *finite* probability spaces. This works very well as many of our examples are from card games and casino games which are examined using finite probability spaces.

While this book is about finite probability spaces, the material on permutations and combinations leads naturally into Stirling's formula for approximating $n!$. To do justice to this and to prepare the reader for future study of continuous distributions, I present the relevant calculus material. This does not assume you have studied calculus previously. So, even if you have a thorough understanding of permutations and combinations, you will find something new to you in this chapter.

Many courses and books on probability theory start with considering a bag of n red balls and m black balls, and then discuss putting your hand into the bag and drawing out, say, three balls and discussing the probability of your having drawn out two red balls and one black ball. In my opinion, this misses the point. Probability is a practical subject. The problems that need to be solved in real life are problems in words. So I make no apology for my examples being rather wordy.

As I said, the aim is to present the material rigorously. Therefore at an early stage we state axioms for a probability space. To do this we need to introduce σ-algebras, and this is indeed what we do in the very first section. While this material may be new to you, it is not difficult.

I have gone to some considerable effort to provide a very detailed index. So if you do not know or cannot remember what something means do not hesitate to look it up in the index at the back.

This book is prepared using LATEX. I have put in a substantial number of hyperlinks which are useful when reading the book on a computer, tablet, or phone.

Acknowledgments

The author owes an enormous debt of gratitude to Professor Ian D. Macdonald who, at the University of Queensland, supervised in 1967/1968 his first research project on varieties of topological groups and free topological groups, which later became the topic of the research for his PhD thesis at Flinders University. The author has continued doing research on related topics for over half a century.

Igor Kluvanek

The author's PhD study was supervised by Professor Igor Kluvanek (1931–1993), who introduced him to free compact abelian groups, to socialism, to slivovitz, and who influenced the author's approach to teaching mathematics as did the author's lecturers: Dr Sheila Oates Williams, Professor Anne Penfold Street (1932–2016), Professor Rudolf Výborný (1928–2016), Professor Des Nicholls, Professor Clive Selwyn Davis (1916–2019), Kenneth Capell, Michael Patrick O'Donnell (1928–1976), Leo Esmond Howard, Brian Lindsay Adkins, Dr Vincent Gerald Michael Hart, Professor Stephen Lipton, Alan Stuart Jones (1939–2023), Dr Anthony McLean Watts, and Dr Keith Robert Mathews who supervised the author's first tutoring at the University of Queensland. Particular thanks go to Dr John Belward (1938–2021) who devoted many hours to assisting the author when a young student to transition to studying more advanced mathematics.

Hanna Neumann

The author's love of mathematics was enhanced in 1963–1964 by the exceptional teacher, Professor Graham Jones, at Cavendish Road State High School, and later enriched by two mathematicians who had a profound influence on Australian pure mathematics: Professors Hanna Neumann and Bernhard H. Neumann of the Australian National University.

B.H. Neumann

Preface

The author has had interesting and very useful conversations over the last 35 years on research and on the teaching of mathematics with his coauthor and friend Professor Karl Heinrich Hofmann of the Technical University of Darmstadt in Germany and Tulane Unversity in New Orleans, USA. Over the last 25 years Hofmann and Morris have coauthored 5 editions of the 1,000 page reference work "The Structure of Compact Groups" [3] and 2 editions of the advanced monograph "The Structure of Pro-Lie Groups" [4] and 36 research papers.

Morris and Hofmann

As a student the author was taught probability theory by Professor Des Nicholls, Henry Finucan, and Professor Stephen Lipton. At that time he used the textbooks [1,2,5].

The author's appreciation of probability and statistics in later years was enhanced by the Egyptian-born Australian statisticians Professor Joseph Gani (1924–2016) and Professor Abraham Michael Hasofer (1927–2010).

Thanks to Graziano Aglietti, Zia Akbar, Anders Jørgensen, Petr Kazil, Carl C. Jung, David Simpson, Yan Yablonovskiy, my co-author and friend Dr Bevan Thompson, Professor Terry Mills, and anonymous referees who provided useful remarks about the presentation and contents of this book. The author thanks his Springer mathematics editor Donna Chernyk and the production supervisor Gomathi Mohanarangan. However, any and all faults with this book that remain are entirely mine.

References

1. Feller, W.: An Introduction to Probability Theory and its Applications. Wiley, New York (1968)
2. Feller, W.: An Introduction to Probability Theory and its Applications, vol. 2, 2nd edn. Wiley, New York (1971)
3. Hofmann, K.H., Morris, S.A.: The Structure of Compact Groups: A Primer for the Student–A Handbook for the Expert, 5th edn. De Gruyter, Berlin (2023). https://doi.org/10.1515/9783111172606
4. Hofmann, K.H., Morris, S.A.: The Structure of Pro-Lie Groups, 2nd edn. EMS Press, Zurich (2023). https://ems.press/books/etm/270
5. Lindley, D.: Introduction to Probability and Statistics from a Bayesian Viewpoint Part 1. Cambridge University Press, Cambridge (1980)

Credit for Images

- Igor Kluvanek. Private collection of Sidney A. Morris.
- Hanna Neumann: By Konrad Jacobs, Erlangen - The Oberwolfach Photo Collection, CC BY-SA 2.0 de,
  https://commons.wikimedia.org/w/index.php?curid=6739429
- B. H. Neumann: By Konrad Jacobs, Erlangen, Copyright is MFO - Mathematisches Forschungsinstitut Oberwolfach,
  http://owpdb.mfo.de/detail?photo$_$id=3003,CCBY-SA2.0de,
  https://commons.wikimedia.org/w/index.php?curid=12341341
- Morris and Hofmann: Private collection of Sidney A. Morris.

Melbourne, VIC, AustraliaSidney A. Morris
June, 2024

Introduction

Misuse of Statistics

We have all heard of Florence Nightingale (1820–1910). However, few realize that she provides an extraordinary example of the power of statistics. She was appalled by the unsanitary conditions she experienced in British army hospitals during the Crimean war. She collected and analysed data on these conditions and presented her findings to Parliament. According to [3], mortality in the Scutari-Crimean Hospital half a year after she arrived had remarkably dropped from 42.7 to 2.2%. Florence Nightingale had the advantage of being both a subject expert, in nursing, and a statistician.

Unfortunately in most cases the person reporting experimental results is either a subject expert with little or no understanding of statistics or a statistics expert with little or no understanding of the subject. This often results in misuse of statistics. The authors of [18] give the following definition of *Misuse of Statistics*: Using numbers in such a manner that—either by intent, or through ignorance or carelessness—the conclusions are unjustified or incorrect. For discussion and examples, see [18, 13, 10, 17].

Interpretation of Data: Bullet Holes

Correct interpretation of data is very important but it is not necessarily easy.

Abraham Wald (1902–1950), a Hungarian Jewish mathematician, worked on decision theory and econometrics. He did his PhD at the University of Vienna, graduating in 1931. Even though he was brilliant, he was unable to win a university position in Austria due to discrimination against Jews. When Germany annexed Austria in 1938, he accepted an invitation to migrate to the USA. During World War II, many US planes returned from missions with bullet hole damage. The military decided that certain areas of the planes needed to be strengthened and chose somewhat naturally those areas with the most bullet holes.

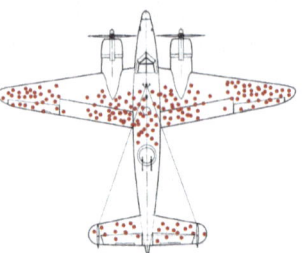

Abraham Wald was part of the Statistical Research Group at Columbia University. He applied the theory of Survival Bias which is the error of concentrating on the things that made it through the selection process and overlooking those that did not. Wald noted that the military considered only the aircraft that survived their mission. Planes that had been shot down were not available for assessment. The bullet holes in the returning aircraft, then, represented areas where a bomber could take damage and still fly well enough to return safely to base.

Wald

Thus Wald proposed that the military reinforce instead the areas where returning aircraft did not have bullet holes, since they were the areas that, if hit, would cause the plane to be lost. Wald deduced that it was the engines which were particularly vulnerable: if these were hit the plane went down and didn't return to base. The military listened and armoured the engine not the wings and tail.

We see the value of stepping back and thinking.

Gambling

One of the first mathematicians who studied probability was Gerolamo Cardano (1501–1576). He used his understanding of probability in gambling and wrote a book called *Book on Games of Chance*.

There is no doubt that games of chance provide a rich environment in which to study probability. This book presents a myriad of examples on permutations and combinations using various games.

Today gambling (including sports gambling) is a giant industry. In Australia, for example, on average, each person (adult and child) expends $1,000 per year gambling. This is a staggering amount. Unfortunately, many people become addicted to gambling, which too often results in their losing not only all the money and possessions they own but also their family and friends and sometimes they are jailed as they steal or conduct fraud to support their addiction.

I wish to make it abundantly clear that this book does not teach you strategies to gamble successfully.

For a history of gambling in (what is now) the United States see [2].

If you need help with a gambling addiction, you may care to contact
http://www.gamblersanonymous.org

History of Probability

The reader should be aware that probability is a subject which developed over hundreds of years. This book makes no attempt to do justice to that history. While we try to include a few biographical details of key players, the interested reader is directed to [1, 4, 5, 8, 9, 12, 18].

Vanderbilt University Library Collection of Books on Card Games and Gaming

According to Nancy Dwyer writing on the Vanderbilt University site "The Library has acquired The George Clulow collection, one of the greatest collections of books about card games, games of chance, playing cards, and chess in the world. This collection, owned since 1903 by the U.S. Playing Card Co., complements the library's Parkhurst and Jane Wood Bridge Collection of Books and Periodicals and has the additional connection to the university in the fact that Vanderbilt's former chairman of the Board of Trust, Harold Stirling Vanderbilt (1884–1970), was the inventor of contract bridge. ...

Included in the collection are books and manuscripts from the 15th to 20th century dealing with the economics, mathematics and social consequences of gaming, as well as the legal ramifications, the art of playing card design, theological diatribes, literary treatments and the mysteries and science of games of chance.

Along with nearly every edition of Hoyle's Game of Whist and plenty of strategy books on poker, bridge, patience, quadrille, skat, and various Italian, French, German and English games, come first editions of literary works in which gaming or gambling play a part, such as Alexander Pope's The Rape of the Lock (1st edition), Swift's The Gambler (1777), and Thackeray's Orphan of Pimilco (1876).

The collection has been called "one of the most complete and scholarly that has ever been gathered together" (Hargrave 1930)."

Solutions to Exercises

There are many exercises in this book. Only by working through a good number of exercises will you master the course. I have not provided solutions to the exercises, and I have no intention of doing so. It is my opinion that there are enough worked examples and proofs within the text itself, that it is not necessary to provide answers to exercises—indeed it is probably undesirable to do so.

Harder exercises are indicated by an *.

However, if you really feel the need to see worked solutions, then look at the book [7] which has one thousand exercises in probability with their solutions. and the book [11] which has nearly 800 problems with comprehensive solutions.

The book *Problems in Probability* [14] by my colleague Professor Terry Mills teaches a substantial amount of probability through presenting problems in probability. And he provides solutions to all of the problems. He says that many of the problems tend to be quite long verging on small research projects.

Software

I first taught computer programming 50 years ago and probability and statistics 45 years ago. At all times since there has been argument about which software languages and packages are best to use. I served as Chair of the Professors of Computers Science and Heads of Departments of Computer Science in Australia for two years and saw the religious fervour with which these were debated by experts. So I know that there is little likelihood that there will be universal agreement. Today there are very powerful and expensive statistics software packages and there are programming languages which have millions of users and advocates. In this book I have chosen to use R and WolframAlpha. These are free and powerful. I do not suggest that these are the best choices, but rather that they have served my purposes very well.

I have made no attempt to teach the reader to use the software package R. However, I include a significant number of examples in R to allow the reader to improve their knowledge. Early in the book little expertise in R is required in order to understand the examples. As the book proceeds, the examples become a little more complicated, particularly as regards drawing graphs. I am not aiming at the cleverest or most elegant solution to drawing the graphs, but rather ways that the reader can learn to use. A very useful book on drawing graphs in R is [13]. (The Lillis book can be purchased from the usual booksellers or it can be subscribed to at https://www.packtpub.com/product/r-graph-essentials/9781783554553, currently for $5 for 5 months. I am not associated with this book in any way.)

Thanks to Wikipedia

I must acknowledge that in many places I have drawn on information in Wikipedia, https://www.wikipedia.org/. In particular, Wikipedia made my job very much easier by identifying which images are in the Public Domain. I say thank you to Wikipedia. And I encourage all readers to consider making a financial contribution to Wikipedia as I have done. It is a very valuable resource.

Genesis of Calculating Chance: Card and Casino Games

In December 2019 I attended the Australian Mathematical Society Annual Meeting in Canberra. While there, I had a discussion with Dr Loretta Bartolini, then Mathematics Editor of the publisher Springer Nature about a second edition of my book *Abstract Algebra and Famous Impossibilities*, which has since appeared [15]. We strayed in our conversation into talking about books on probability theory. Soon after that, I decided to write a book on probability. The COVID-19 pandemic in 2020 caused me to focus my efforts on writing such a book which, in due course, became this book.

References

1. Bell, E.T.: Men of Mathematics: The Lives and Adventures of Great Mathematicians from Zeno to Poincaré. Simon and Schuster, New York (1986)
2. Chaeftz, H.: Play the Devil: A History of Gambling in the United States from 1492 to 1955, Potters Publishers, New York (1960)
3. Cohen, B.: Florence nightingale. Sci Am **250**(3), 128–137 (1984)
4. David, F.N.: Games, Gods and Gambling. Hafner Publishing, New York (1962)
5. Fischer, H.: A History of the Central Limit Theorem: From Classical to Modern Probability Theory. Springer, New York (2011)
6. Gigerenzer, G.: Calculated Risks: How to Know When Numbers Deceive You. Simon & Schuster, New York (2003)
7. Grimmett, G.R., Stirzaker, D.R.: One Thousand Exercises in Probability. Oxford University Press. New York (2001)
9. Hald, A.: A History of Mathematical Statistics from 1750 to 1930. Wiley, New York (1998)
8. Hald, A.: A History of Probability and Statistics and Their Applications Before 1750. Wiley, New Jersey (2003)
10. Huff, D.: How to Lie with Statistics. W. W. Norton, New York (1993)
11. Kelly, M.W., Donnelly, B.: Humongous Book of Statistics Problems (Nearly 800 Problems with Comprehensive Solutions. Alpha, Dubai (2010)
11. Kotz, S., Johnson, N.L.: Breakthroughs in Statistics Volume I: Foundations and Basic Theory. Springer, New York (1992)

12. Krämer, W., Gigerenzer, G.: How to confuse with statistics or: the use and misuse of conditional probabilities. Statist. Sci. **20**(3), 223–230 (2005). https://projecteuclid.org/euclid.ss/1124891288
13. Lillis, D.A.: R Graph Essentials. Packt Publishing, Birmingham (2014)
14. Mills, T.M.: Problems in Probability, 2nd edn. World Scientific, Singapore (2014)
15. Morris, S.A., Jones, A. Pearson, K.R.: Abstract Algebra and Famous Impossibilities: Squaring the circle; Doubling the Cube; Trisecting an Angle; and Solving Quintic Equations, 2nd edn. Springer Nature, Switzerland (2022)
16. Ross, J.: Misuse of statistics in social sciences. Nature **318**, 514 (1985). https://doi.org/10.1038/318514a0
17. Spirer, H.F., Spirer, L., Jaffe, A.J.: Misused Statistics, 2nd edn. Dekker, New York (1988)
18. Todhunter, I.: A History of the Mathematical Theory of Probability: From the Time of Pascal to that of Laplace. Chelsea Publishing, New York (1949)

Credits for Images

- Florence Nightingale. Public Domain.
- Plane damage. This file is licensed under the Creative Commons Attribution-Share Alike 4.0 International license.
  https://commons.wikimedia.org/wiki/File:Survivorship-bias.png
- Abraham Wald. Public Domain.

Contents

Preface ix

Introduction xiii

1 **Finite Probability Spaces and Examples** 1
  1.1 Introduction and σ-Algebras 1
  Problems 6
  1.2 The Event Space and the Probability Space 8
      Random Variable, Expected Value, and Variance 10
      Monopoly: Going to Jail and Getting Out of Jail 13
      Gambler's Fallacy 14
      Playing Cards 15
      Tarot Playing Cards 16
      Roulette 17
      Autosomal Recessive Diseases 19
      Sic Bo 21
      Martingale Betting System 23
      Fibonacci Betting System 23
      Setting Winnings and Losses Bounds 24
      Waiting Times for a Bus 24
      Hazard 25
      Craps 27
      Chuck-a-Luck; Crown and Anchor 29
      Blackjack 30
      Billionaire Gambling 35
      Horseracing Odds 35
      Macao 37
      Baccarat 38
      Ballot Box Problem 40

		Weather Forecasting and Chaos Theory	42
		Simpson's Paradox	45
	Problems		47
		Klondike Solitaire	47
	1.3	Credit for Images	50
	References		53
2	**Permutations and Combinations**		55
	2.1	Permutations and Combinations	55
	2.2	Card Games	57
		The 52 card deck	58
		The Standard 52-card deck	58
		The Origin of the Sandwich	59
		Cribbage, Noddy, and Costly Colours	59
		Sir John Suckling: Inventor of Cribbage	60
		Muggins in the English Language	60
		Cribbage in Literature, Movies, TV, and Video Games	60
		Self-proclaimed Cribbage Capital of the World—Nelson, Montana	61
		Cribbage Game Rules as Explained by ChatGPT	61
		Poker	65
		Combination Lock	65
		Selection With and Without Replacement	65
		Bluffing in Poker	70
		Bridge	71
		Bidding in Bridge	73
		Euchre	74
		Rummy	75
	2.3	Mahjong	76
		Learning Mahjong and its role in Modern American Culture	79
		American Mahjong or Mah Jongg	80
	2.4	Lotto	80
	2.5	Voltaire Discovers a Flaw in French Lotto and Becomes a Millionaire	81
	2.6	Bingo	82
	Problems		83
		Dead Man's Hand in Poker	83
		Samuel Pepys Dice Problem	84
		Dreidels	86
		Two-Up	88
		Birthday Problem	89
	2.7	Credit for Images	90
	References		91

3 Conditional Probability ... 93
3.1 Conditional Probability ... 93
Intellectual Honesty ... 93
Viruses ... 94
Cells, Chromosomes, and DNA ... 95
Polymerase Chain Reaction (PCR) ... 97
Virus Testing ... 97
Kolmogorov's Definition of Conditional Probability ... 99
Bayes' Theorem ... 101
Bertrand's Box Problem ... 102
Monty Hall Problem ... 103
Life Expectancy ... 105
Bayesian Theory ... 105
John Maynard Keynes and Probability ... 106
Problems ... 107
Martin Gardner's Three Prisoners Problem ... 108
3.2 Credit for Images ... 110
References ... 111

4 Stirling's Approximation Formula and Improvements ... 113
4.1 Stirling's Approximation Formula ... 113
4.2 Sequences and Limits ... 116
L'Hôpital's Rule ... 117
Asymptotic ... 119
4.3 Series ... 119
4.4 Infinite Integrals ... 120
Trigonometry ... 122
4.5 Evaluating the Gaussian Integral $\int_{-\infty}^{\infty} e^{-x^2}\, dx$ ... 123
4.6 The Standard Normal Distribution ... 124
Limit Comparison Test for Infinite Integrals ... 125
4.7 The Gamma Function ... 126
Integration by Parts ... 127
4.8 The Gamma Distribution ... 128
Taylor Series ... 131
Differentiable Functions ... 132
4.9 Proofs from The Book ... 133
4.10 Laplace Extension of Stirling's Formula ... 134
4.11 Improvements on Stirling's Formula ... 138
4.12 Male Births ... 143
4.13 The Basel Problem: Evaluating $\zeta(2) = \sum_{n=1}^{\infty} \frac{1}{n^2}$ ... 144

4.14 Probability that Two Randomly Chosen Natural Numbers Are
Coprime 149
Existence of an Infinite Number of Prime Numbers 150
Proof by Contradiction 151
4.15 Principle of Inclusion and Exclusion........................... 156
Carefree Couples 158
Problems 158
De Montmort's Matching Problem 160
4.16 Credit for Images... 161
References 162

Index 165

Chapter 1
Finite Probability Spaces and Examples

Abstract

In this first chapter probability theory is placed on a firm foundation. Then a variety of interesting and entertaining examples are introduced and investigated. These include French and American *roulette*; *dice games* such as craps, hazard, chuck-a-luck, crown and anchor, and sic-bo; and *card games* such as blackjack, macao, baccarat, and solitaire. Beyond the realms of card and casino games, topics ranging from the Martingale betting system and the Fibonacci betting system, setting winnings and losses bounds, the "optimum" strategy for blackjack, the gamblers' fallacy, inherited diseases, monopoly, weather forecasting, chaos theory, horseracing odds, elections, your intuition, and paradoxes are examined. While we should expect that in each and every casino game the house has a statistical advantage, in this chapter we shed light on which games offer players a good chance.

1.1 Introduction and σ-Algebras

David Hilbert (1862–1943), arguably the pre-eminent mathematician of his time, wanted to place probability theory on a firm foundation.

Just as vector space theory, Euclidean geometry, group theory, topology, and graph theory each begin with a set of axioms, he sought a set of axioms as a starting point for probability theory.

At the second International Congress of Mathematicians, held in Paris in 1900, Hilbert posed 23 questions which were to be influential on research in the twentieth century. His sixth problem included the task of providing a set of axioms for probability theory.

Hilbert

It was solved in 1933 by the outstanding Russian mathematician Andrej Nikolajewitsch Kolmogorov (1903–1987), who made significant contributions to our understanding of classical mechanics, computational complexity, information theory, logic,

Kolmogorov

topology, as well as probability theory. He also solved Hilbert's 13th problem [19].

In this chapter we present his axioms for probability spaces. First we record some notation.

Notation.
(i) the set of all natural numbers $\{1, 2, \ldots, n, \ldots\}$ is denoted by $\mathbb{N}$;
(ii) the set of all integers is denoted by $\mathbb{Z}$;
(iii) the set $\{\frac{p}{q}; p, q \in \mathbb{Z}, q \neq 0\}$ of all rational numbers is denoted by $\mathbb{Q}$;
(iv) the set of all real numbers is denoted by $\mathbb{R}$;
(v) the set of all complex numbers is denoted by $\mathbb{C}$;
(vi) the set $\mathbb{R} \setminus \mathbb{Q}$ of all irrational real numbers is denoted by $\mathbb{P}$;
(vii) the set of all real numbers x satisfying $a \leq x \leq b$, for $a, b \in \mathbb{R}$ with $a \leq b$, is denoted by $[a, b]$
(viii) the set of all real numbers x satisfying $a < x < b$, for $a, b \in \mathbb{R}$ with $a < b$, is denoted by (a, b);
(ix) the set of all real numbers x satisfying $x \geq a$, for $a \in \mathbb{R}$, is denoted by $[a, \infty)$;
(x) the set of all real numbers x satisfying $x \leq a$, for $a \in \mathbb{R}$, is denoted by $(-\infty, a]$.

While the reader is no doubt keen to get on with discussing probability itself, we proceed methodically. The first step is to ensure that any gaps in knowledge are filled.

In this section we define what we mean by a σ-algebra on a set X. We shall see that a σ-algebra is defined to be a set of subsets of X having four simple properties. To describe these properties, we will need to be familiar with properties of the union and intersection of a finite or infinite number of sets and how these behave with respect to taking complements.

Definition 1.1 if S is a subset of a set E, then the *complement* of S in E, denoted by $E \setminus S$ or S', is the set of all members of E which are not in S; that is, $E \setminus S = S' = \{x : x \in E, x \notin S\}$.

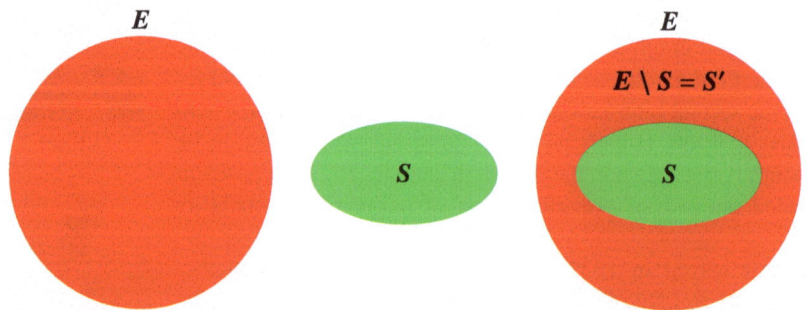

Example 1.1 The complement of $\mathbb{Q}$ in $\mathbb{R}$ is $\mathbb{P}$ and the complement of $(0, 1)$ in $\mathbb{R}$ is $(-\infty, 0] \cup [1, \infty)$.

1.1 Introduction and σ-Algebras

Definition 1.2 If I is any finite or infinite index set and each $S_i, i \in I$, is a subset of a set E, then $\bigcup_{i \in I} S_i$ denotes the subset of E which consists of all $x \in E$ which belong to at least one of the S_i.
Similarly, $\bigcap_{i \in I} S_i$ is the subset of E which consists of those $x \in E$ such that $x \in S_i$, for all $i \in I$.
If $I = \mathbb{N} = \{1, 2, \ldots, n, \ldots\}$, then we write $\bigcup_{i \in I} S_i$ as $\bigcup_{i=1}^{\infty} S_i$ and $\bigcap_{i \in I} S_i$ as $\bigcap_{i=1}^{\infty} S_i$.

Example 1.2 The complement of $\mathbb{Z}$ in $\mathbb{R}$ is given by

$$\mathbb{R} \setminus \mathbb{Z} = \bigcup_{n=0}^{\infty}(n, n+1) \;\cup\; \bigcup_{n=0}^{\infty}(-n-1, -n).$$

To work comfortably with complements, one needs to be aware of *De Morgan's Laws*, proved by the English mathematician Augustus De Morgan (1806–1871).

If A and B are subsets of a set X, then

(i) $X \setminus (A \cap B) = (X \setminus A) \cup (X \setminus B)$; and
(ii) $X \setminus (A \cup B) = (X \setminus A) \cap (X \setminus B)$.

De Morgan

More generally, we have:
If I is any finite or infinite set and $S_i, i \in I$, are subsets of a set X, then

(a) $X \setminus \left(\bigcap_{i \in I} S_i\right) = \bigcup_{i \in I}(X \setminus S_i)$; and
(b) $X \setminus \left(\bigcup_{i \in I} S_i\right) = \bigcap_{i \in I}(X \setminus S_i)$.

We shall prove (a) and leave the proof of (b) as an exercise.

Proof of (a).

> To prove two sets J and K are equal, we show that if x is in J, then x is in K, and if $x \in K$, then $x \in J$.

Assume $x \in X \setminus \left(\bigcap_{i \in I} S_i\right)$
$\implies x \in X$ and $x \notin \bigcap_{i \in I} S_i$
$\implies x \in X$ and there is an $i_1 \in I$ such that $x \notin S_{i_1}$
$\implies x \in X \setminus S_{i_1}$
$\implies x \in \bigcup_{i \in I}(X \setminus S_i)$.

Conversely, assume $x \in \bigcup_{i \in I}(X \setminus S_i)$
$\implies$ there is an $i_2 \in I$ such that $x \in X \setminus S_{i_2}$
$\implies$ since $X \setminus \left(\bigcap_{i \in I} S_i\right) \supseteq X \setminus S_{i_2}$, we have that $x \in X \setminus \left(\bigcap_{i \in I} S_i\right)$.

So we have proved (a). □

Definition 1.3 Let X and Y be sets and f a mapping of X into Y. If S is a subset of Y, then the *inverse image* of the set S is the subset of X consisting of all those x such that $f(x) \in Y$ and is denoted by $f^{-1}(S)$; that is, $f^{-1}(S) = \{x \in X : f(x) \in S\}$.

> **! Attention**
>
> If X and Y are sets, then a function $f : X \to Y$ is said to have an *inverse function* $g : Y \to X$ if $f(g(y)) = y$, for all $y \in Y$ and $g(f(x)) = x$, for all $x \in X$. For $f : X \to Y$ to have an inverse function it is necessary and sufficient that f is one-to-one and *onto*.
> Recall that f is *one-to-one* (also known as *injective*) if for any $x_1 \in X$ and $x_2 \in X$ with $f(x_1) = f(x_2)$, then $x_1 = x_2$. Also recall f is said to be *onto* (also known as *surjective*) if for each $y \in Y$, there is at least one $x \in X$ such that $f(x) = y$.
> So we observe that $f_1 : \mathbb{R} \to \mathbb{R}$ given by $f_1(x) = |x|$ is not a one-to one function and so has no inverse function. Similarly $f_2 : \mathbb{R} \to \mathbb{R}$ given by $f_2(x) = x^4$ is not onto and so has no inverse function. The function $f_3 : \mathbb{R} \to \mathbb{R}$ given by $f_3(x) = \sin x$ is neither one-to-one nor onto and so has no inverse function.
> Unfortunately when $f : X \to Y$ is one-to-one and onto and does therefore have an inverse function $g : Y \to X$, some books write this inverse function g as f^{-1}. This is misleading.
>
> By Definition 1.3, if $f : X \to Y$ is any function from X into Y, the inverse image $f^{-1}(S)$ of each subset S of Y exists and is a subset of X.
> By contrast the function f has an inverse function $g : Y \to X$ if and only if it is one-to-one and onto.

Example 1.3

(i) Let (i) $f_1 : \mathbb{R} \to \mathbb{R}$ be given by $f_1(x) = |x|$, for all $x \in \mathbb{R}$. Then

$$f_1^{-1}([1,2]) = [-2,-1] \cup [1,2] \text{ and } f_1^{-1}(\{2\}) = \{-2,2\} \text{ and } f_1^{-1}([-1,\infty)) = \mathbb{R};$$

(ii) Let $f_2 : \mathbb{R} \to \mathbb{R}$ be given by $f(x) = x^4$, for all $x \in \mathbb{R}$. Then

$$f_2^{-1}(\{81\}) = \{-3,3\} \text{ and } f_2^{-1}(\mathbb{R}) = \mathbb{R} \text{ and } f_2^{-1}([-1,1]) = [-1,1];$$

(iii) Let $f_3 : \mathbb{R} \to \mathbb{R}$ be given by $f_3(x) = \sin x$, for all $x \in \mathbb{R}$. Then

$$f_3^{-1}(\mathbb{R}) = \mathbb{R} \text{ and } f_3^{-1}([-10,10]) = \mathbb{R} \text{ and } f_3^{-1}([-1,1]) = \mathbb{R}.$$

Greek Alphabet. It is helpful to know a few Greek letters and their names.

(i) **sigma**: lower case σ; upper case Σ;
(ii) **gamma**: lower case γ; upper case Γ;
(iii) **delta**: lower case δ; upper case Δ;

1.1 Introduction and σ-Algebras

(iv) **phi**: lower case ϕ; upper case Φ;
(iv) **omega**: lower case ω; upper case Ω;
(v) **psi**: lower case ψ; upper case Ψ;
(vi) **eta**: η;
(vii) **mu**: μ;
(viii) **epsilon**: ε;
(ix) **xi**: ξ.

There is a mathematical symbol borrowed from the Danish alphabet, namely, $\emptyset$, which denotes the empty set, and is not to be confused with the Greek letter ϕ.

Notation. For a finite set S, we denote by $|S|$ the number of elements of S.

Definition 1.4 Let Ω be any set and Σ a set of subsets of Ω. Then Σ is said to be a σ-algebra on Ω if it satisfies the following four conditions:

(i) $\Omega \in \Sigma$;
(ii) if the subset S of Ω is in Σ, then its complement $S' = \Omega \setminus S \in \Sigma$;
(iii) if $S_1, S_2, \ldots, S_n, \ldots$ are in Σ, then $\bigcap_{n=1}^{\infty} S_n \in \Sigma$;
(iv) if $S_1, S_2, \ldots, S_n, \ldots$ are in Σ, then $\bigcup_{n=1}^{\infty} S_n \in \Sigma$.

Proposition 1.1 *If Σ is a σ-algebra on a set Ω, then*

(i) $\emptyset \in \Sigma$;
(ii) *if $S_1, S_2, \ldots, S_n \in \Sigma$, then $S_1 \cap S_2 \cap \cdots \cap S_n \in \Sigma$;*
(iii) *if $S_1, S_2, \ldots, S_n \in \Sigma$, then $S_1 \cup S_2 \cup \cdots \cup S_n \in \Sigma$;*
(iv) *If $S \in \Sigma$ and $A \in \Sigma$ with $A \subset S$, then $S \setminus A \in \Sigma$.*

Proof. Exercise. $\square$

Example 1.4

(i) Let Ω be any set. The set $\mathcal{P}(\Omega)$ of all subsets of Ω (known as the *power set of* Ω) is a σ-algebra on Ω. The power set of a set Ω is often denoted by 2^{Ω}.
(ii) Let Ω be any set and $\Sigma_0 = \{\Omega, \emptyset\}$. Then Σ_0 is a σ-algebra on Ω. This σ-algebra, Σ_0, is called the *trivial σ-algebra on Ω*.
(iii) Let Ω be the set $\mathbb{N}$ and Σ be the set of all subsets of $\mathbb{N}$ containing only even integers. Then Σ is not a σ-algebra on $\mathbb{N}$ since, for example, the set $\{2, 4, 8\}$ is in Σ but its complement is not in Σ.
(iv) Let Ω be the set $\mathbb{R}$ and Σ consist of all finite subsets of $\mathbb{R}$. Then Σ is not a σ-algebra on $\mathbb{R}$ since, for example, all singleton sets $\{n\}$, for $n \in \mathbb{N}$, are in Σ but their union $\bigcup_{n=1}^{\infty} \{n\} = \mathbb{N}$ is not in Σ. So condition (iv) is not satisfied.
(v) Let Ω be the set $\{1, 2, 3, 4\}$ and Σ consist of $\{1\}$, $\{1, 3\}$, $\{2, 3, 4\}$, $\{2, 4\}$, $\emptyset$, and $\{1, 2, 3, 4\}$. Then Σ is not a σ-algebra on Ω as $\{1\} \cup \{2, 4\}$ is not in Σ and so condition (iii) is not satisfied.

The next proposition tells us that property (iv) in Definition 1.4 is redundant; that is, it is already implied by the other three properties.

Proposition 1.2 *Let Ω be any set and Σ a set of subsets of Ω. Then Σ is a σ-algebra if it satisfies the following three conditions:*

(i) $\Omega \in \Sigma$;
(ii) *if the subset S of Ω is in Σ, then its complement $S' = X \setminus S \in \Sigma$;*
(iii) *if $S_1, S_2, \ldots, S_n, \ldots$ are in Σ, then $\bigcap_{n=1}^{\infty} S_n \in \Sigma$.*

Proof. Exercise. □

Proposition 1.3 *Let X and Y be any sets, $f : X \to Y$ any function, and let Σ be a σ-algebra on Y. Define the set Λ of subsets of X as follows:*
$$S \in \Lambda \text{ if and only if } S = f^{-1}(E), \text{ where } E \in \Sigma.$$
Then Λ is a σ-algebra on X.
(In other words, the inverse image of a σ-algebra is a σ-algebra.)

Proof. Exercise. □

Problems

> The great scholar and teacher, Rabbi Adin Steinsaltz (1937–2020), asked a young teacher "Would you eat a sandwich after I had chewed it?" Of course not replied the teacher.
> With this in mind, I do not want to give you problems that I have already chewed on and given you solutions. I want you to chew on them afresh.

1.1 Prove Proposition 1.1. [Hint. Given $S_1, S_2, \ldots S_n$, for some $n \in \mathbb{N}$, put $S_m = \emptyset$ for every $m \in \mathbb{N}$ with $m \geq n + 1$.]

1.2 Let Ω be any set and Σ a set of subsets of Ω which has the property that for all S_1, S_2 with $S_1 \in \Sigma$ and $S_2 \in \Sigma$, we have $S_1 \cup S_2 \in \Sigma$. Prove using mathematical induction that Σ satisfies property (iii) of Proposition 1.1.

1.3 Prove Proposition 1.2

1.4 Prove (b) of De Morgan's laws.

1.5 Prove Proposition 1.3.

1.6 Let Ω be a set and let Σ_1 and Σ_2 be σ-algebras on the set Ω. Define Σ_3 as follows:
$$\text{for a subset } S \text{ of } \Omega, S \in \Sigma_3 \iff S \in \Sigma_1 \text{ and } S \in \Sigma_2.$$

Prove that Σ_3 is a σ-algebra on Ω.
Thus *the intersection of any two σ-algebras on Ω is a σ-algebra on Ω.*

1.1 Introduction and σ-Algebras

1.7 Let $\Omega = \{1, 2, 3\}$. Find σ-algebras Σ_1 and Σ_2 on Ω such that Σ_3 is not a σ-algebra on Ω, where

$$\text{for a subset } S \text{ of } \Omega, \ S \in \Sigma_3 \iff S \in \Sigma_1 \text{ or } S \in \Sigma_2.$$

So the union of two σ-algebras is not necessarily a σ-algebra.

1.8 Let Σ be a σ-algebra on the set $\mathbb{R}$. Prove the following:

(i) If for each $a \in \mathbb{R}$ and $b \in \mathbb{R}$ with $a \leq b$, the open interval $(a, b) \in \Sigma$, then each of the intervals $[c, d], [c, d), (c, \infty), (-\infty, d), [c, \infty), (-\infty, d]$ is in Σ. [Hint: The closed interval $[0, 1]$ can be expressed as an infinite intersection of open intervals as follows: $[0, 1] = \bigcap_{n=1}^{\infty} (-\frac{1}{n}, 1 + \frac{1}{n}).$]

(ii) If for each $a \in \mathbb{R}$ and $b \in \mathbb{R}$ with $a \leq b$, the closed interval $[a, b] \in \Sigma$, then each of the intervals $[c, d], [c, d), (c, \infty), (-\infty, d), [c, \infty), (-\infty, d]$ is in Σ.

(iii) If for each $a \in \mathbb{R}$ and $b \in \mathbb{R}$ with $a \leq b$, the interval $[a, b) \in \Sigma$, then each of the intervals $[c, d], [c, d), (c, \infty), (-\infty, d), [c, \infty), (-\infty, d]$ is in Σ.

(iv) If for each $a \in \mathbb{R}$, $(a, \infty) \in \Sigma$, then $(c, d] \in \mathbb{R}$ for all $c, d \in \mathbb{R}$. [Hint: Use Definition 1.4(ii).]

(v) If for each $a, b \in \mathbb{Q}$, $[a, b] \in \Sigma$, then $(c, d) \in \Sigma$, for all $c, d \in \mathbb{R}$.

1.9 Let Σ be a σ-algebra on $\mathbb{Z}$ such that $\{n\} \in \Sigma$, for every $n \in \mathbb{Z}$. Prove that every subset S of $\mathbb{Z}$ is in Σ.

1.10 Let Σ be a σ-algebra on $\mathbb{R}$ such that for every $a \in \mathbb{R}$, the singleton set $\{a\} \in \Sigma$. Prove that $\mathbb{Z} \in \Sigma$ and every subset of $\mathbb{Z}$ is in Σ.
(It can also be proved that both $\mathbb{Q}$ and $\mathbb{P}$ are in Σ, once one knows that $\mathbb{Q}$ is a countable set.)

1.11 Let Ω be a non-empty set and T a set of subsets of Ω. Then T is said to be a *topology* on Ω if it satisfies the four conditions (a)–(d):

(a) $\Omega \in T$;
(b) $\emptyset \in T$;
(c) for each $n \in \mathbb{N}$, $S_1 \in T, S_2 \in T, \ldots, S_n \in T$ implies $S_1 \cap S_2 \cap \cdots \cap S_n \in T$;
(d) if I is any set and each $S_i \in T$, for $i \in I$, then $\bigcup_{i \in I} S_i \in T$.

Prove the following:

(i) if Ω is a finite set and Σ is a σ-algebra on X, then Σ is also a topology on Ω;
(ii) if $\Omega = \{1, 2, 3, 4\}$, then there is a topology T on X such that T is not a σ-algebra on Ω;

If t is a topology on a non-empty set Ω, the smallest σ-algebra Σ on Ω such that $\Sigma \supset t$ is called the *Borel σ-algebra*. The members of a Borel σ-algebra are called *Borel sets*. Borel sets are named after the French mathematician Félix Édouard Justin Émile Borel (1871–1956). Borel sets play an important role in the area of mathematics called measure theory [1, 5, 16, 24]

1.2 The Event Space and the Probability Space

In this section we formally define the notion of probability space and give several examples using familiar games.

The first book about games of chance *Liber de ludo aleae* (Book on Games of Chance), was written by the Italian mathematician and physician Gerolamo Cardano (1501–1576) in the 1560s, but not published until the seventeenth century. In fact Cardano supported himself through medical school on winnings from gambling using his understanding of probability.

Cardano

Definition 1.5 Let Ω be a (finite or infinite) set, Σ a σ-algebra on the set Ω, and $I = \{1, 2, \ldots, n\}$ for some $n \in \mathbb{N}$ or $I = \mathbb{N}$. A *probability measure* or a *probability distribution* P is a function from Σ to the closed unit interval $[0, 1]$ with the following properties:

(i) $P(\emptyset) = 0$;
(ii) $P(\Omega) = 1$;
(iii) if sets $S_i \in \Sigma$, for $i \in I$, are such that $S_i \cap S_j = \emptyset$, for each $i, j \in I$ with $i \neq j$, then
$$P\left(\bigcup_{i \in I} S_i\right) = \sum_{i \in I} P(S_i).$$

In this context, the set Ω is said to be the *sample space* and the set Σ is said to be the *event space*. The triple (Ω, Σ, P) is said to be a *probability space*. If $A \in \Sigma$, then the number $P(A) \in [0, 1]$ is said to be the *probability of the event A* (or the *probability that A occurs*).

For clarity, we observe that in Definition 1.5

- $P(S_1 \cup S_2 \cup \cdots \cup S_n) = P(S_1) + P(S_2) + \cdots + P(S_n)$
- $P\left(\bigcup_{i=1}^{\infty} S_i\right) = \sum_{i=1}^{\infty} P(S_i).$

Notation. If $A, B \in \Sigma$, then $A \cap B$ is denoted by AB. If $A_1, \ldots, A_n \in \Sigma$, for $n \in \mathbb{N}$, then $A_1 \cap A_2 \cdots \cap A_n$ is denoted by $A_1 A_2 \ldots A_n$. If $A_1, \ldots, A_n \cdots \in \Sigma$, for $n \in \mathbb{N}$, then $\bigcap_{n=1}^{\infty} A_n = A_1 \cap A_2 \cdots \cap A_n \ldots$ is denoted by $A_1 A_2 \ldots A_n \ldots$.

Definition 1.6 Let Ω be any set, I any index set and $S_i, i \in I$, subsets of Ω. The sets $S_i, i \in I$, are said to be *pairwise disjoint* if for each $i, j \in I$ with $i \neq j$, $S_i \cap S_j = \emptyset$. If Ω is a sample space, Σ an event space on Ω, and $S_i \in \Sigma$, for $i \in I$, then the events S_i are said to be *mutually exclusive* if for each $i, j \in I$, $i \neq j$, $S_i \cap S_j = \emptyset$.

Before looking at some simple examples, we observe a useful proposition for some finite sample spaces.

1.2 The Event Space and the Probability Space

Proposition 1.4 *Let $\Omega = \{\omega_1, \omega_2, \ldots, \omega_n\}$, for some $n \in \mathbb{N}$, be a finite sample space, the event space $\Sigma = 2^\Omega$, and P a probability measure. If*

$$P(\{\omega_1\}) = P(\{\omega_2\}) = \cdots = P(\{\omega_n\}),$$

then for each set $A \in \Sigma$, $P(A) = \frac{|A|}{n}$.

Proof. Exercise. □

Example 1.5 I have a coin with a head (H) on one side and a tail (H) on the other side.
(On a 2020 Australian 20 cent coin, the head side is that of Queen Elizabeth II and the tail side is actually a platypus, a uniquely Australian animal.)

I throw the coin up, and it lands with either the head facing up or the tail facing up.
The sample space $\Omega = \{H, T\}$.
For the event space, Σ, I choose the power set.
So $\Sigma = \{\emptyset, \{H, T\}, \{H\}, \{T\}\}$.
The coin is said to be *fair coin* for a probability measure P if $P(\{H\}) = P(\{T\}) = \frac{1}{2}$, so heads and tails are equally likely.
Using Proposition 1.4, we therefore have, for each $A \in \Sigma$, $P(A) = \frac{|A|}{2}$, since $n = 2$; that is,
$P(\emptyset) = 0$, $P(\{H, T\}) = 1$, $P(\{H\}) = \frac{1}{2}$, and $P(\{T\}) = \frac{1}{2}$.

Fair Die

Example 1.6 I have a die with six sides.

> We note that the plural of die is dice

Each of the six sides has 1, 2, 3, 4, 5, or 6 dots with no two sides having the same number of dots. I throw the die in the air, and it lands with one side up. I am interested in knowing whether the side that is up has an even number of dots.

The sample space Ω is $\{1, 2, 3, 4, 5, 6\}$. For the event space, I choose the power set. So $\Sigma = \mathcal{P}(\{1, 2, 3, 4, 5, 6\})$. Thus Σ has $2^6 = 64$ members.

The die is said to be a *fair die* for probability measure P if $P(\{1\}) = P(\{2\}) = P(\{3\}) = P(\{4\}) = P(\{5\}) = P(\{6\}) = \frac{1}{6}$.

So for each $A \in \Sigma$, again using Proposition 1.4,

$$P(A) = \frac{|A|}{|\Omega|} = \frac{|A|}{6}.$$

I want the number of dots on the side facing up to be an even number; that is, 2, 4, or 6. Thus, I seek $P(\{2, 4, 6\})$.

So $P(\{2, 4, 6\}) = \frac{3}{6} = \frac{1}{2}$.

Casinos generally offer several different games of chance such as baccarat, blackjack, roulette, craps, pontoon, poker, slot machines, etc. With each game, there is a chance of winning and a bigger chance of losing. You may want to choose a game where your chances of winning are best. To do this, the notion of expected value is introduced. This concept was first defined explicitly in 1814 in a publication by the French polymath Pierre-Simon Laplace (1749–1827) (known to many as the inventor of the Laplace transform [9]). You may think of expected value as a kind of weighted average. (Of course there is dispute about who first came up with the notion of expected value—see [15].)

Laplace

Random Variable, Expected Value, and Variance

Definition 1.7 Let (Ω, Σ, P) be a probability space. Further, let Ω be a finite set $\{\omega_1, \omega_2, \ldots, \omega_n\}$ and Σ be the power set 2^Ω of Ω.

A function $X : \Omega \to \mathbb{R}$ is said to be a *random variable*.

So X has a finite number of possible values $X(\omega_1), X(\omega_2), \ldots, X(\omega_n)$, and these occur with probability $P(\{\omega_1\}), P(\{\omega_2\}), \ldots, P(\{\omega_n\})$, respectively.

The *expected value of X* (also known as the *expectation of X*) is denoted $E[X]$ and is defined by

1.2 The Event Space and the Probability Space

$$\mu = E[X] = \sum_{i=1}^{n} X(\omega_i) P(\{\omega_i\}).$$

The *variance of* X is denoted by $\mathrm{Var}(X)$ and defined by

$$\mathrm{Var}(X) = E[(X-\mu)^2] = E[X^2] - (E[X])^2.$$

The *standard deviation of* X is denoted by σ and defined to be $\sqrt{\mathrm{Var}(X)}$, the square root of the variance.

Remark 1.1 The definitions of random variable and expected value in Definition 1.7 are specifically for the case that the sample space Ω is finite and the event space Σ is its power set 2^Ω.

Example 1.7 In Example 1.5 of tossing a coin, we define a random variable $X : \{H, T\} \to \mathbb{R}$ by $X(T) = 5$ and $X(H) = 6$, then the expected value is given by $E[X] = 5 \times \frac{1}{2} + 6 \times \frac{1}{2} = 5.5$.

[Using the software package R: $5*1/2 + 6*1/2$]

Example 1.8 In Example 1.6 of throwing a die, we define the random variable $X : \{1, 2, 3, 4, 5, 6\} \to \mathbb{R}$ by $X(1) = 1, X(2) = 2, \ldots, X(6) = 6$.
So the expected value $\mu = E[X] = \sum_{i=1}^{6} (i \times \frac{1}{6}) = \frac{1}{6} + \frac{2}{6} + \cdots + \frac{6}{6} = 3.5$.

To calculate the variance and standard deviation, observe that X^2 has the values $1, 4, 9, 16, 25, 36$.

So we see that $E[X^2] = \frac{1}{6}(1 + 4 + 9 + 16 + 25 + 36) = \frac{91}{6}$.

$$\mathrm{Var}(X) = E[X^2] - (E[X])^2 = \frac{91}{6} - \frac{49}{4} = \frac{35}{12} = 2.916\ldots.$$

The standard deviation $\sigma = \sqrt{\mathrm{Var}(X)} = \sqrt{\frac{35}{12}} = 1.707\ldots.$

We can check our calculation using R. Caution must be used as there is a difference between sample variance and variance. R uses *var* to calculate sample variance whereas we are calculating variance.

```
y<- c(1,2,3,4,5,6)
mean(y)
f1<-function(x) {
z<- (mean(y^2)- (mean(y))^2)
return(z)        } #f1 is the variance
f1(z)
sqrt(f1(z)) # this is the standard deviation
```

In statistics it is very important to distinguish between the population and a sample of the population. The population consists of all members. A sample is a part of the population which allows us to make inferences about the population as a whole. The sample size may be quite small compared to the population size.

The variance and standard deviation in Definition 1.7 are the *population variance* and the *population standard deviation*. The *sample variance* and *sample standard deviation* are used to estimate the population variance and the population standard deviation.

The population variance as we have seen is given by $\sigma^2 = \dfrac{\sum_{i=1}^{n}(x_i - \mu)^2}{n}$.

The sample variance is defined by $s^2 = \dfrac{\sum_{i=1}^{n}(x_i - \text{mean}\{x_1, x_2, \ldots, x_n\})^2}{n-1}$.

In 1918 the English statistician and geneticist Sir Ronald Aylmer Fisher (1890–1962) introduced the notion of variance. Fisher became Professor of Genetics at Cambridge University in 1943 and was knighted in 1952. He was described by Danish statistician Anders Hjorth Hald (1913–2007) in [14] as "a genius who almost single-handedly created the foundations for modern statistical science" and by the American statistician Bradley Efron (born 1938) in [11] as "the single most important figure in twentieth century statistics".

Fisher

Fisher was a prominent opponent of Bayesian statistics but in 1950 was the first to use the term "Bayesian".

Unfortunately Fisher is also remembered as having controversial views on race as indicated by his being founding Chairman in 1911 of the University of Cambridge Eugenics Society. Fisher held a favourable view of eugenics even after World War II. Fisher wrote about the Nazi party that it "sincerely wished to benefit the German racial stock, especially by the elimination of manifest defectives" and that he would give "his support to such a movement". Such a statement today would be regarded as racist.

Definition 1.8 Let (Ω, Σ, P) be a probability space and $X : \Omega \to \mathbb{R}$ and $Y : \Omega \to \mathbb{R}$ random variables. If

$$P(X = x, Y = y) = P(X = x) \cdot P(Y = y), \text{ for all } x, y \in \mathbb{R}$$

then X and Y are said to be *independent random variables*.

Example 1.9 In Greek mythology we learn of Achilles whose mother held him by his heel and dipped him in the river Styx. As a result Achilles became invulnerable

1.2 The Event Space and the Probability Space

except for the part of his heel by which his mother had held him. This is the proverbial "Achilles heel".

For the purposes of this example, we shall assume that there are 65 archers who each aim a deadly arrow at Achilles heel and that these arrows are independent random variables with a probability of hitting their target of $\frac{1}{100}$. Is it more likely that Achilles will be fatally wounded than not?

> Our intuition might tell us that if each arrow has a 1% chance of hitting his heel, then 65 arrows have more than 50% chance of hitting his heel. However, we are dealing with independent random variables. So if one arrow has a probability of $\frac{99}{100}$ of not hitting his heel, then by Definition 1.8 two arrows have probability $\left(\frac{99}{100}\right)^2$ of not hitting his heel. So we obtain the calculation below which shows the answer to our question is no.

The probability that he will not be fatally wounded is, by Definition 1.8,

$$\left(\frac{99}{100}\right)^{65} = 0.5203405\ldots.$$

Monopoly: Going to Jail and Getting Out of Jail

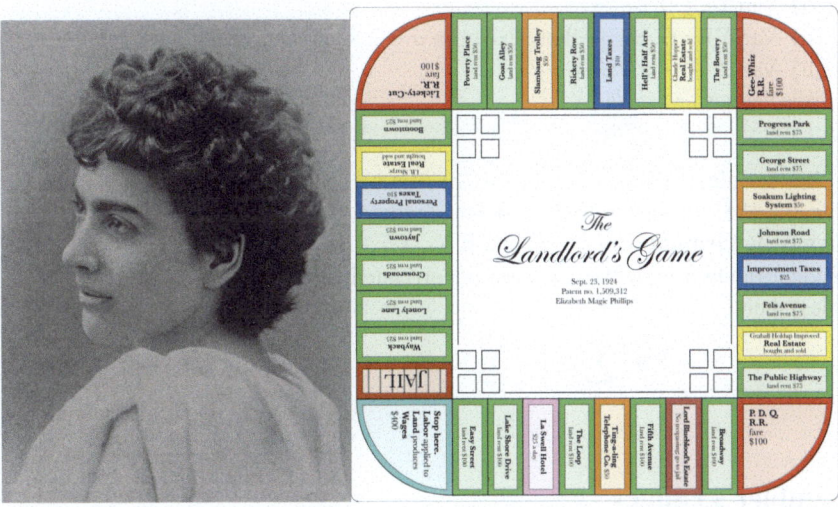

Example 1.10 The game of Monopoly™ is probably the most popular board game in history having sold hundreds of millions and being available in dozens of languages.

Its history can be traced back to 1903 when the poet, author, engineer, comedian, actress, and feminist Elizabeth J. Magie Phillips (née Magie) (1866–1948) invented and patented in 1904 the board game she called *The Landlord's Game*. She intended it to be an educational tool to show how wrong it was for ownership of land to be concentrated in monopolies. She patented the game again in 1923.

Monopoly is not a game of pure chance but rather a game of chance and strategy. However, one aspect is pure chance. In Monopoly you roll two dice. If you roll a double, you roll the dice again. If you roll a double again, you roll the dice a third time. If this third time you also roll a double, you Go to Jail. Rolling the dice three times are three independent events. So the probability of three doubles is $(\frac{1}{6})^3 = 0.004626296\ldots$. So the probability of going to jail by this method is very low.

Let us now turn to Getting Out of Jail. This can be achieved by rolling a double on any one of three rolls of the dice. We shall calculate the probability of NOT rolling a double on any of the three rolls of the dice. As these are independent events, the probability is $(\frac{5}{6})^3$. So the probability of rolling a double on at least one of the three rolls of the dice is $1 - (\frac{5}{6})^3 = 0.421296\ldots$. So the probability of Getting Out of Jail by this method is not too bad.

Definition 1.9 Let (Ω, Σ, P) be a probability space and $X : \Omega \to \mathbb{R}$ and $Y : \Omega \to \mathbb{R}$ random variables. If $E[X \cdot Y] = E]X) \cdot E[Y]$, then X and Y are said to be *uncorrelated random variables*.

We note that uncorrelated does not imply independent.

We note that for X any random variable and a any (constant) real number:

(i) $\text{Var}(X) \geq 0$;
(ii) $\text{Var}(X) = 0 \iff$ there exists an $a \in \mathbb{R}$ such that $P(X = a) = 1$;
(iii) $\text{Var}(X + a) = \text{Var}(X)$;
(iv) $\text{Var}(aX) = a^2 \text{Var}(X)$.

Irénée-Jules Bienaymé (1796–1878) was a French statistician. In 1853 he discovered what has become known as the **Bienaymé formula** which says that if $X_1, X_2, \ldots, X_n$ are independent random variables (or even uncorrelated random variables) then

$$\text{Var}\left(\sum_{i=1}^n X_i\right) = \sum_{i=1}^n \text{Var}(X_i).$$

Bienaymé

Gambler's Fallacy

Before we leave the topic of independent random variables, we should mention what is known as *gamblers' fallacy*. Many gamblers tend to believe that past events can

1.2 The Event Space and the Probability Space

influence future events, even when they are independent. For example, if I throw a fair coin and three consecutive times it lands on heads, then they believe that it is more likely to land on tails on the next throw. If it is indeed a fair coin, then it has a 50% chance of landing on tails and a 50% chance of landing on heads next time. This misunderstanding is called gambler's fallacy. In my example, it is pretty harmless, but in a high-stakes bet in a casino, it is not harmless.

Playing Cards

Example 1.11 I have a deck of playing cards. There are 52 cards in the deck: ace,2,3,4,5,6,7,8,9,10,jack,queen,king in each of the four suits: spades, hearts, diamonds, and clubs as pictured below. For brevity we refer to them as

AS, 2S,3S,4S,5S,6S,7S,8S,9S,10S,JS,QS,KS
AH, 2H,3H,4H,5H,6H,7H,8H,9H,10H,JH,QH,KH
AD, 2D,3D,4D,5D,6D,7D,8D,9D,10D,JD,QD,KD
AC, 2C,3C,4C,5C,6C,7C,8C,9C,10C,JC,QC,KC

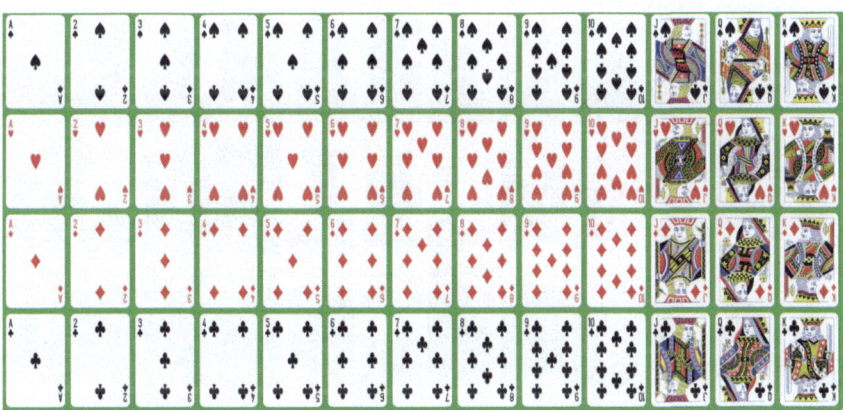

I select one card at random from the deck.
I am interested in whether that card is an ace.
The sample space Ω has 52 members.
I choose the event space Σ to be the power set 2^Ω. So Σ has 2^{52} members.
By saying that I choose a card at random, I mean that the probability measure P is such that for each $\omega \in \Omega$, $P(\{\omega\}) = \frac{1}{52}$.
So by Proposition 1.4, for any event $E \in \Sigma$, $P(E) = \frac{|E|}{52}$.
For $E = \{AS, AH, AD, AC\}$, then $P(E) = \frac{4}{52} = \frac{1}{13}$.
The probability of selecting an ace at random is therefore $\frac{1}{13}$.

Tarot Playing Cards

In the English-speaking world when one hears of *Tarot cards* one immediately thinks of the occult, fortune-telling, cartomancy. However, tarot cards have been used in European countries such as Italy, France, Germany, and Austria since the fifteenth century for playing card games. Tarot cards were introduced about 1420 for the purpose of playing card games, and indeed it was in these games that two notions, fundamental to the game of bridge, first appeared, namely, that of trumps and bidding. For a detailed history and description of the many games played with tarot cards, see [10]. Tarot card games are played with a 78 card deck having four ordinary suits and an additional suit of trumps.

There is dispute about the origin of tarot cards as used for occult purposes. A protestant pastor Antoine Court de Gébelin (1725–1725) asserted that tarot was a repository of ancient wisdom dating back thousands of years. The British philosopher

1.2 The Event Space and the Probability Space

Michael Dummett (1925–2011) scoffed at such a history of tarot. The author of this book does not wish to enter this controversy. We shall focus only on tarot cards as used in playing games.

Roulette

Example 1.12 Wikipedia describes the casino game of roulette as follows: "Roulette is a casino game named after the French word meaning little wheel. In the game, players may choose to place bets on either a single number, various groupings of numbers, the colours red or black, whether the number is odd or even, or if the numbers are high (19–36) or low (1–18).

To determine the winning number, a croupier spins a wheel in one direction, then spins a ball in the opposite direction around a tilted circular track running around the outer edge of the wheel. The ball eventually loses momentum, passes through an area of deflectors, and falls onto the wheel and into one of 37 (single zero French/European style roulette) or 38 (double zero American style roulette) coloured and numbered pockets on the wheel. The winnings are then paid to anyone who has placed a successful bet".

French roulette wheel

The sample space Ω for this French roulette wheel is $\{0, 1, 2, \ldots, 36\}$.
We shall put the event space Σ as the power set, 2^Ω, of the sample space. Then Σ has 2^{37} members.
We shall assume that the roulette wheel is fair, which means that

$$P(\{\omega\}) = \frac{1}{37}, \text{ for each } \omega \in \Omega.$$

Again by Proposition 1.4, if $A \in \Sigma$, then $P(A) = \frac{|A|}{37}$.

A variety of bets is allowed. We mention some.

Bet $1 on a single number.
Let the bet be on the number ω. Then $P(\{\omega\}) = \frac{1}{37}$. If the ball lands in the pocket numbered ω, then the player generally wins $36. If the ball lands on any of the other 36 numbers, the player loses the $1. So the player has probability of $\frac{1}{37}$ of a payout of $36 and a probability of $\frac{36}{37}$ of a payout of $0. We call the payout X_1. Then X_1 is a function from $\{0, 1, 2, \ldots, 36\}$ to $\mathbb{R}$ with $X_1(\omega) = 36$ and $X(x) = 0$ for $x \neq \omega$. Of course X_1 is a random variable. So the expected value $E[X_1]$ of the payout in dollars is

$$\sum_{i=0}^{36} X_1(i) \times \frac{1}{37} = \left(36 \times \frac{1}{37}\right) + 36\left(0 \times \frac{1}{37}\right) = \frac{36}{37}.$$

So the player can expect a loss of $\$\frac{1}{37}$ on each $1 bet; that is about a 2.7% loss.

An alternative to a bet on a single number is a bet on red or black, that is, a bet that the ball lands in the pocket of one of the 18 red numbers or one of the 18 black numbers.

On the French roulette wheel, there is also one green pocket labelled 0.

If the player bets $1 on red, we see that $P(\text{red}) = \frac{18}{37}$. Generally if the ball lands on red, the player receives $2, and if it lands on black or green, the player loses the $1.

So the payout random variable $X_2 : \{0, 1, 2, \ldots, 36\} \to \mathbb{R}$ with

$$X_2(\text{any number in a red pocket}) = 2, \text{ and}$$

$$X_2(\text{any number in a black or green pocket}) = 0.$$

So the expected value $E[X_2]$ of the payout is $(2 \times \frac{18}{37}) + (0 \times \frac{19}{37}) = \frac{36}{37}$.
The player can expect a loss of $\$\frac{1}{37}$ on each $1 bet, that is, again about a 2.7% loss. Note the expected loss on this bet is exactly the same as a bet on a single number.

Similarly, the player can bet low or high, that is, bet, that the ball will land in the pocket of one of the low numbers $1, 2, \ldots, 18$ *or one of the high numbers* $19, 20, \ldots, 36$.

We see that, as in the last case, the probability that the ball lands on a low number (or a high number) is $\frac{18}{37}$. For a winning bet the player generally receives $2. So once again the player can expect to get back from the $1 bet $\$\frac{36}{37}$. So the expected loss is the same as previously.

The last case we shall look at is a bet on four numbers that meet at one corner, such as $10, 11, 13, 14$.

The probability that the ball lands on one of these four numbers is $\frac{4}{37}$ and the payout is generally $9. And a similar analysis shows that the expected loss is the same as for the previously considered bets.

Before discussing the American roulette wheel, we mention one difference sometimes between the French roulette and the European roulette as regards the inclusion in the French case of the La Partage rule. With the La Partage rule, if the ball lands on zero, then the player receives half of an even-money bet such as red/black or odd/even back without any winnings. Not always do casinos offer the La Partage rule even for French roulette.

We conclude by looking at the *American roulette wheel*. In contrast with the French or European roulette wheel which has 18 numbers in red pockets, 18 numbers in black pockets, and one number 0 in a green pocket, the more common American roulette wheel has 18 numbers in red pockets, 18 numbers in black pockets, plus 0 in a green pocket and double zero 00 in a green pocket. This changes the odds in the favour of the house (i.e. the casino). Then $P(\{x\}) = \frac{1}{38}$. Generally the payout for a single bet on the number ω is $36. So the expected loss on a $1 bet is $\$(\frac{36}{38} - 1) = \$\frac{1}{19}$ or about 5.3% of the bet.

1.2 The Event Space and the Probability Space

Before continuing on casino games, let us have a change of pace and consider health related examples.

Autosomal Recessive Diseases

> Before presenting the next example, I record the fact that *I have no medical qualifications*. When I discuss a medical topic, you should understand that my purpose is to look at it from the perspective of probability theory. If you want to know medical facts, then this book is not the place to look. The presentation is my understanding as a non-medically qualified person.

Example 1.13 According to the World Health Organization (WHO): Monogenic diseases result from modifications in a single gene occurring in all cells of the body. Though relatively rare, they affect millions of people worldwide. Scientists currently estimate that over 10,000 human diseases are known to be monogenic. Pure genetic diseases are caused by a single error in a single gene in the human DNA. The single-gene diseases can be classified into three categories: dominant, recessive, and X-linked.
All human beings have two copies of each gene– one copy on each side of the chromosome pair. Recessive diseases are monogenic disorders that occur due to damages in both copies. Monogenic diseases are responsible for a heavy loss of life. About 1% of all babies in the world have a single gene disease at birth.

As examples we mention three autosomal recessive diseases listed by the World Health Organization. (Autosomal means the gene in question is one of the 22 pairs (44 in all) of non-sex genes.)

  (i) *Sickle-cell anaemia (SCA):* sickle-cell anaemia is a blood-related disorder that affects the haemoglobin molecule and causes the entire blood cell to change shape under stressed conditions. In sickle-cell anaemia, the haemoglobin molecule is defective. Normal red blood cells live about 120 days in the bloodstream, but sickled red cells die after about 10–20 days. Because they cannot be replaced fast enough, the blood is chronically short of red blood cells, leading to a condition commonly referred to as anaemia. The disease occurs in about 1 in every 500 African-American births and 1 in every 1,000–1,400 Hispanic-American births. About 2 million Americans, or 1 in 12 African Americans, carry the sickle cell gene.
  (ii) *Cystic fibrosis (CF):* cystic fibrosis is a genetic disorder that affects the respiratory, digestive, and reproductive systems involving the production of abnormally thick mucus linings in the lungs and can lead to fatal lung infections. The disease can also result in various obstructions of the pancreas, hindering digestion. In the USA the incidence of CF is reported to be 1 in every 3,500 births.

(iii) *Tay-Sachs disease (TSD):* Tay-Sachs disease is a fatal genetic disorder, named after the British ophthalmologist Warren Tay (1843–1927) and the American neurologist Bernard Sachs (1858–1954), in which harmful quantities of a fatty substance called Ganglioside GM2 accumulate in the nerve cells in the brain. The frequency of the condition is much higher in Ashkenazi Jews of Eastern European origin than in others. Approximately 1 in every 27 Jews in the USA is a carrier of the TSD gene. Amongst Jews of Sephardic origin and in the general, non-Jewish population, the carrier rate is about 1 in 250. The carrier rate amongst Irish Americans is about one in 50.

In autosomal recessive inheritance, both copies of the gene in each cell have mutations. The parents of an individual with an autosomal recessive condition each carry at least one copy of the mutated gene, but they typically do not show signs and symptoms of the condition. Autosomal recessive disorders are typically not seen in every generation of an affected family.

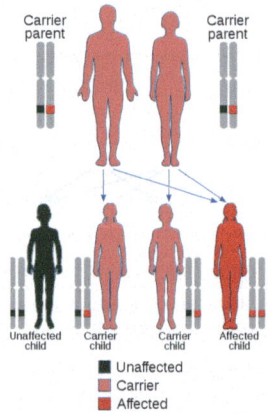

Autosomal recessive

Let us now look at these from the perspective of probability theory. We see that a person may have zero, one, or two copies of the mutated genes. If they have two copies of the mutated gene, then they have the disease. If they have one copy of the mutated gene, they are a carrier.

(a) If the person has exactly one copy of the mutated gene, then they do not have the disease, but are a carrier. A child of theirs can have the disease or can be a carrier or not be a carrier.
(b) If the person has zero copies of the mutated gene, then they do not have the disease and are not a carrier. A child of theirs cannot have the disease but can be a carrier.
(c) If the person has two copies of the mutated gene, then they have the disease and a child of theirs will either have the disease or be a carrier.

Let us consider the case where each parent does not have the autosomal recessive disease but is a carrier. Then each has a mutated gene (MG) and a non-mutated gene (NG). For convenience we shall write the gene inherited from the father first and [MG NG] or [NG MG].

The sample space Ω for the child's two relevant genes:

$$\Omega = \{[\text{MG MG}], [\text{MG NG}], [\text{NG MG}], [\text{NG NG}]\}.$$

We let $\Sigma = 2^\Omega$ be the event space.

If we regard which of the two genes MG and NG the child receives from the father as equally probable and which of the two genes MG and NG the child receives from the mother as equally probable, then we have the probability space $(\Omega, 2^\Omega, P)$ where (for the child)

1.2 The Event Space and the Probability Space

$$P([\text{MG MG}]) = P([\text{MG NG}]) = P([\{\text{NG MG}]) = P([\text{NG NG}]) = \frac{1}{4}.$$

The child has the autosomal recessive disease only in the case [MG MG], and this has probability $\frac{1}{4}$. The child is a carrier of the autosomal recessive disease in the cases [MG {NG}] and [NG MG] and $P(\{[\text{MG NG}], [\text{NG MG}]\}) = \frac{1}{4} + \frac{1}{4} = \frac{1}{2}$.

Sic Bo

Sic Bo, also known as Tai Sai and Dai Siu, originated in China. It is a casino game, played with three dice, probably introduced into the USA by Chinese immigrants. Wherever there is a significant Chinese community, it is likely that local casinos offer Sic Bo. In this game, three dice are rolled and players win or lose on the result of that one roll of the three dice.

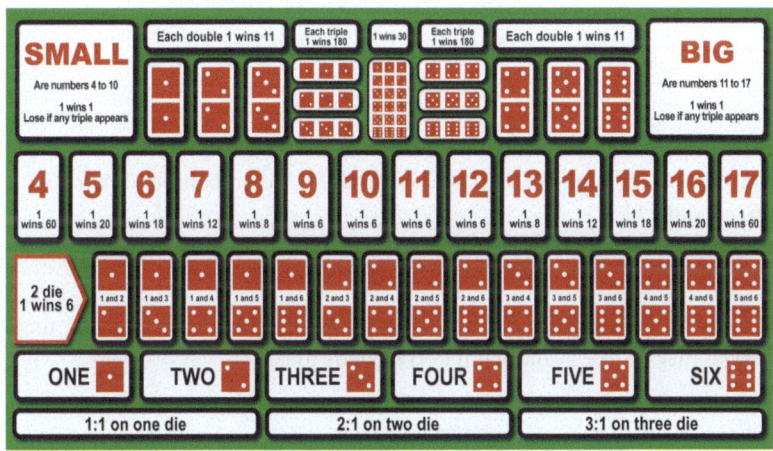

Layout of Sic Bo Table

As indicated by the above table layout, there is a variety of possible bets. However we shall focus on some simple examples.

Players place their bets. Then the dealer picks up a small chest containing the three dice which they shake and then open and reveal one number on each of the three dice.

The first bet we consider is called SMALL, that is, the sum of the three numbers showing on the dice is 4,5,6,7,8,9,or 10 with the exception that the numbers are not equal.

It is easy to see that each of the possible sums of the numbers showing on the three dice can usually be obtained in more than one way. In all there are 216 different ways. The table below, constructed by writing down all the possibilities, indicates the number of different ways each number can be obtained.

Sum	Number of ways
3	1
4	3
5	6
6	10
7	15
8	21
9	25
10	27
11	27
12	25
13	21
14	15
15	10
16	6
17	3
18	1

So the number of dice throws which result in a SMALL which is not a triple is $3 + 6 + (10 - 1) + 15 + 21 + (25 - 1) + 27 = 105$. So the probability of a SMALL is $105/216 = 0.486111\ldots$.

Typically the casino will offer *odds* 1:1 on a SMALL.

If the odds of an event are $a : b$ for some positive real numbers a and b, then the probability of that event that it corresponds to is

$$\frac{1}{\frac{a}{b}+1} = \frac{b}{a+b}.$$

So odds of $1 : 1$ correspond to a probability of 0.5. Odds of $30 : 1$ correspond to a probability of $0.032258\ldots$.

Now odds of $30 : 1$ mean if you bet \$1 and you win, then you receive \$30 plus you get back your \$1 bet back.

If the odds are $1 : 1$ and you win, then you receive \$1 plus your bet of \$1 back.

So the casino is offering odds of $1 : 1$ corresponding to a probability of 0.5 on a SMALL. So in the long-term the percentage profit to the casino is $(0.5 - 0.486111\ldots)/(0.5)$ or about 2.78%.

The second bet we consider is BIG, that is, the sum of the numbers is $11, 12, 13, 14, 15, 16, 17$ with the exception that the numbers are not equal.

The arithmetic for BIG is exactly the same as for SMALL. The profit for the casino is the same as for SMALL.

1.2 The Event Space and the Probability Space

The casino will typically offer odds of 30 : 1 for a *triple*. There are 6 triples of the 216 possibilities. So the probability of a triple is $\frac{6}{216} = 0.02777\ldots$. We have seen that the odds 30 : 1 are equivalent to a probability of $0.032258\ldots$. So the casino percentage profit in the long term is $(0.032258\cdots - 0.02777\ldots)/0.032258\ldots$ which is about 13.9%.

Of course there are many other ways to bet on the outcome of rolling the three dice.

Sometimes casinos make the environment more exciting for players by having all winning bets on the table light up immediately the dice have been rolled.

Martingale Betting System

We have probably all seen it. When somebody loses a bet, they say "double or nothing".

The *Martingale betting system* is that of following each losing bet by doubling it. The idea is that by repeatedly doubling the bet, *eventually* you will break even. There are some difficulties with this system.

A gambler does not have infinite funds and so may run out of money to double their losing bet before they win, also they may run out of time to gamble (e.g. the casino may close), or there may be a limit on the size of bets allowed (e.g. by the casino).

There is another serious shortcoming of the Martingale system.

You are guaranteed to at least break even only if you stop betting once you have won or have broken even. Most gamblers cannot so easily stop!

Fibonacci Betting System

The Fibonacci numbers, [26], were first described over 2,000 years ago in Sanskrit poetry. However they are named after the Italian mathematician Leonardo of Pisa (about 1,170–about 1,250), who is better known as Fibonacci, who introduced these numbers in his 1,202 book *Liber Abaci*. This sequence of numbers is 0, 1, 1, 2, 3, 5, 8, 13, 21, 34, 55, 89, 144, ..., where each number after the second is the sum of the previous two numbers.

We shall focus on the numbers 1, 2, 3, 5, 8, 13, 21, 34, 55, 89, 144, ...

The Fibonacci betting system, [25], is best applied in games with a chance of winning close to 50%.

The idea is simple: each time you lose a bet, move one number up the Fibonacci sequence. Each time you win, move two numbers down the Fibonacci sequence.

Let us show this with a simple example:

We start with a bet of $5 and we lose. So we now have a loss of $5 and our next bet is $5 \times 2 = $10.

We lose and so we now have a loss of $15 and our next bet is $5 × 3 = $15.
We lose and so we now have a loss of $30 and our next bet is $5 × 5 = $25.
We lose and so we now have a loss of $55 and our next bet is $5 × 8 = $40.
We lose and so we now have a loss of $95 and our next bet is $5 × 13 = $65.
We win and so we now have a loss of $30 and our next bet is $5 × 5 = $25.
We win and so we now have a loss of $5 and our next bet is $5 × 2 = $10.

So when we lose, we increase the amount we bet, but not by as much as we would in the Martingale betting system. Unfortunately, there is no obvious point to stop betting in the Fibonacci betting system.

Also, and most importantly, neither the Martingale betting system nor the Fibonacci betting system does anything to remove the statistical advantage that the House may have.

Setting Winnings and Losses Bounds

It is a very smart idea, in advance of any day of betting, to decide

(1) I will stop betting if my winnings total x dollars;
(2) I will stop betting if my losses total y dollars.

And then stick very firmly to what you have decided.

Let us again have a change of pace and consider a very different problem.

Waiting Time for a Bus

Remark 1.2 Many people regard probability as a hard subject. What, in fact, is the case is that our intuition often misleads us. We shall see this again and again throughout this book. Here is an example from [28].

Ken walks to the bus stop every day after 6am and before 7am and takes the first bus irrespective of whether it is going north or south. The north-bound buses and the south-bound buses run equally often, every 5 minutes. If Ken ignores the bus timetable and simply goes to the bus stop when he is ready, what is the probability that he will take the north-bound bus?

Most people would say that the probability is about $\frac{1}{2}$. That is what our intuition tells us. But like a good lawyer, we should ask for "further and better particulars".

1.2 The Event Space and the Probability Space

Table 1.1 Bus timetable

North bound 6.00am 6.05am 6.10am 6.15am ···
South bound 6.04am 6.09am 6.14am 6.19am ···

If Ken arrives at the bus stop in the intervals (6.00am, 6.04am], (6.05, 6.09], (6.10, 6.14], (6.15, 6.19], (6.20, 6.24], (6.25, 6.29] he will take the South Bound bus. If he arrives in the intervals (6.04, 6.05], (6.09, 6.10], (6.14, 6.15], (6.19, 6.20], (6.24, 6.25], (6.29, 7.00], he will take the north-bound bus. Clearly he has a 4 times greater chance of catching the south-bound bus than the north-bound bus. So our intuition was incorrect.

Hazard

We shall next consider games played with two dice. Just as in Sic Bo, it is easy to see that if we roll the two dice, then each of the possible sums of the numbers showing on the two dice can usually be obtained in more than one way. In all there are 36. The table below, constructed by writing down all the possibilities, indicates the number of different ways each number can be obtained.

Sum	Number of ways	Probability
2	1	1/36
3	2	2/36
4	3	3/36
5	4	4/36
6	5	5/36
7	6	6/36
8	5	5/36
9	4	4/36
10	3	3/36
11	2	2/36
12	1	1/36

Clearly 7 is the easiest number to roll, while 1 and 12 are the hardest numbers to roll.

We shall consider two such games: first the very old and somewhat complicated game of *Hazard* [30] and then its modern derivative casino game, *Craps*. Hazard is an early English game played with two dice; it was mentioned in The *Canterbury Tales* by Geoffrey Chaucer (1340s–1400). In the play Richard III of William Shakespeare (1564–1616), Richard III in his famous "My kingdom for a horse" speech in Act V says: "I have set my life upon a cast, And I will stand the hazard of the die".

Writing in 1674, Charles Cotton (1630–1687), in his book *The Compleat Gamester*, remarks:

Chaucer portrait 1412 by Thomas Hoccleve (1368–1426)

"Hazard is the most bewitching game that is played on the Dice; for when a man begins to play he knows not when to leave off; ... and having once accustomed himself to play at Hazard, he hardly, ever after, minds anything else".

As one might expect, there are various versions of the rules of this ancient game. It is played with two dice by any number of people. One player, called the *shooter*, begins the game. The shooter declares the "target" called the *main* which is any of the numbers 5, 6, 7, 8, or 9. (Contrast this with the game *Craps* where the main is always 7.) Now the other players place a bet on whether the shooter will win or lose. Then the shooter rolls the two dice.

		Roll result		
Main	2 or 3	Main	11	12
5 or 9	Lose	Win	Lose	Lose
6 or 8	Lose	Win	Lose	Win
7	Lose	Win	Win	Lose

The probability that the shooter who chooses the main of 7 wins at the first roll of the dice equals $6/36 + 2/36 = 8/36 = 0.222\ldots$. The probability that this shooter loses on the first roll of the dice is $4/36 = 0.111\ldots$. So the probability that this shooter will need to roll a second time is $1 - 8/36 - 4/36 = 24/36 = 0.666\ldots$.

The probability that the shooter who chooses the main of 5 or 9 wins at the first roll of the dice equals $4/36 = 0.111\ldots$ The probability that this shooter loses on the first roll of the dice is $6/36 = 0.1666\ldots$. So the probability that this shooter will need to roll a second time is $1 - 4/36 - 6/36 = 26/36 = 0.7222\ldots$.

1.2 The Event Space and the Probability Space

The probability that the shooter who chooses the main of 6 or 8 wins at the first roll of the dice equals $5/36 + 1/36 = 0.1666\ldots$. The probability that this shooter loses on the first roll of the dice is $5/36 = 0.1388\ldots$. So the probability that this shooter will need to roll a second time is $1 - 6/36 - 5/36 = 25/36 = 0.6944\ldots$.

We see that it usually happens that the shooter will have to roll the dice at least a second time.

If the shooter rolls neither the main nor 2, 3, 11, or 12, the number that is rolled is called the *chance*. At this point things flip-flop, so to speak. Now the shooter wants to roll anything other than the main, because if the shooter rolls the main they lose. And if the shooter rolls the chance they win. So the shooter continues to roll the dice until they roll the main and lose or the chance and win.

Finally we mention that the British etymologist Michael Quinion (born 1943) suggested that the English phrase *at sixes and sevens* which describes a state of confusion comes from the game of hazard and was used again by Geoffrey Chaucer in his Troilus and Criseyde. Quinion thinks that originally it referred to fives and sixes which were less good choices of the "main" than seven in the game of hazard.

Hazard reached its peak in the seventeenth and eighteenth centuries. In the nineteenth century the rules of hazard were simplified and the game of *craps* was born which is very popular to today, especially in casinos in America.

Craps

Many of us will have seen the James Bond classic 1971 movie *Diamonds are Forever* starring Sean Connery in which he visits a Las Vegas casino and plays craps.

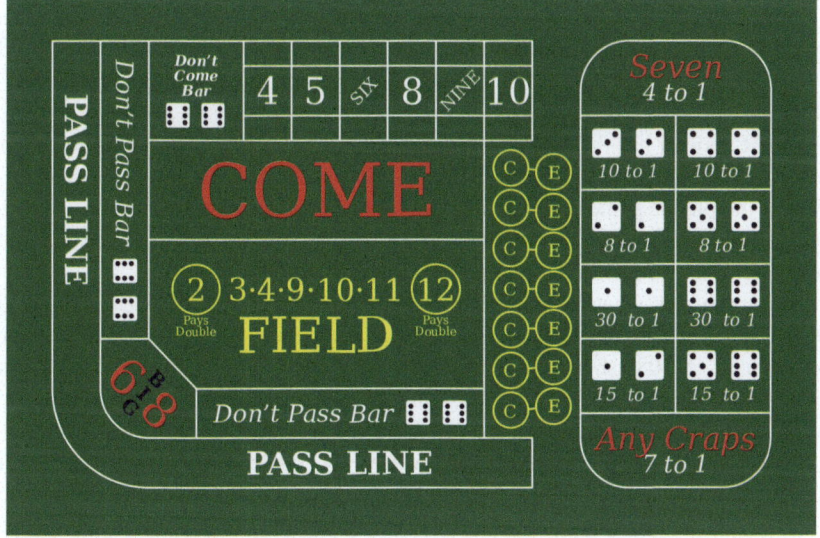

While the Craps Table Layout seems complicated and there are a variety of possible bets, we have seen that craps is a simplified version of the game of *hazard* which we have already described. But for completeness we shall give a simple description of the game of craps. We note that there are many books describing the game of craps, including [4, 8, 17, 20, 30].

The shooter rolls two dice. If the total of the faces showing of the two dice rolled is 7 or 11, which is called a *natural*, then the shooter wins the amount bet. Anyone betting with the shooter, called a *pass* bettor, also wins the same amount. Anyone betting against the shooter, called a *don't pass* bettor, loses the same amount. If a total of 2, 3, or 12 occurs on this first roll, it is called *craps*, then the pass bettors lose and the don't pass bettors win. If the total of the first roll is any of the remaining numbers 4, 5, 6, 8, 9, 10, then this number becomes the pass bettor's *point*. The shooter then continues to roll the dice. If the point is rolled before the 7, the shooter and pass bettors win. Of course if the 7 comes up before the point, then the shooter and the pass bettors lose.

We digress to describe the casino staff. The *Boxman* watches the pile of chips and settles any disputes. The *Stickman* who directs the dice towards the shooter and calls out the result of each throw of the dice and reminds everyone of the point number. There are usually two dealers who stand either side of the Boxman who convert cash to chips and place players' bets on the table. Often there are five dice from which the shooter chooses two. And to roll the dice, the shooter must throw the dice hard enough to bounce off the side of the table to ensure randomness.

We have seen in the analysis of the game of hazard that the probability of the shooter throwing a 7 or 11 and winning immediately is 2/9 and the probability that they throw a 2 or 3 or 11 or 12 and lose immediately is 1/9. So there is a probability of 1/3 that the game ends after one throw.

There is a probability of 2/3 that they throw any other number, and this number then becomes the *point*.

Now everything depends on whether the shooter throws the point before they throw a 7. The probability of throwing a 7 is 1/6. But the probability of throwing the point depends on what the point actually equals. The probability may be as low as 1/36 if the point is 2 or 12 or as high as 5/36 if it is 6 or 8.

So the shooter has the probability of losing on the second throw with probability 1/6 and winning with probability between 1/36 and 5/36.

In principle the game could continue forever with the shooter continuing to throw neither the point nor the 7. However it is not difficult to show that typically the game ends after 3 or 4 throws. (See [4, p. 108].)

You might ask what is the probability that the shooter, and hence any pass line bettor, will win. You have all the data required to do the calculation, but it is tedious. Bollman [4, p. 108] shows the probability is about 0.4929. So the casino which usually pays even money odds of 1:1 has the edge of 1.41%, which is quite low compared, for example, to what we saw for roulette.

As we mentioned earlier, there are a variety of possible bets. One can bet that a specific number will be rolled before the number 7 or a field bet that the next number

1.2 The Event Space and the Probability Space

rolled will be a 2, 3, 4, 9, 10, 11 or, 12. There are *yo* bets that the next number rolled will be a 3—a risky bet with a high payout. These are but a few examples.

Chuck-a-Luck; Crown and Anchor

Chuck-a-Luck also known as *bird cage* or *sweat rag* is played with three dice and can be regarded as a variant of *sic bo* and *grand hazard*. It is essentially the same as the Vietnamese game *Bau cua ca cop*, the Chinese game of *Hoo Hey How*, and the Cambodian game of *Klah Klok*.

Chuck-a-Luck is also the same as the game which originated in the eighteenth century called *Crown and Anchor* played by sailors in the British navy. The dice in Crown and Anchor, rather than having numbers, are marked with the symbols: crown, anchor, diamond, spade, club, and heart.

A similar game called *Langur Burja* is played in Nepal. The game *Crown and Anchor* [22] is mentioned in a 2007 episode set in 1943 about illegal gambling called *Casualties of War* of the magnificent British World War II detective TV series *Foyle's War*.

Chuck-a-Luck is played in the 1974 James Bond movie *The Man with the Golden Gun* starring Roger Moore and referred to in the 1941 Abbott and Costello movie *Hold that Ghost*.

The game Chuck-a-Luck is described in [20]. Usually there are three large dice in a wire cage. The three dice are rolled once. The bettor places a bet, say of $1 on a particular number, for example, the number 6. If the three dice show 6-2-6 then the bettor receives $2 plus his wager of $1 back. Similarly if 1-5-6 are rolled then the bettor would receive $1 plus his wager of $1. So the game seems fair and easy for anyone to understand.

However, if one thinks through the probabilities, then one sees that there are $216 = 6^3$ possible results of rolling the three dice. The number of ways of rolling, for example, two 6s and winning $2 is 6-x-6, x-6-6, and 6-6-x where x in each case can have any one of 5 values. So two 6s can occur in 15 different ways of the total of 216. So the probability of throwing 2 6s is 15/216. The number of ways of rolling three 6s is 1 and so has probability 1/216. The number of ways of rolling one 6 is 75 and so has probability 75/216. So the probability of rolling at least one 6 is $(15 + 1 + 75)/216 = 91/216$.

So if the bettor rolls precisely one 6, which he does with probability 75/216, he wins $1. So if the bettor rolls precisely two 6s, which he does with probability 15/216, he wins $2. So if the bettor rolls precisely three 6s, which he does with probability 1/216, he wins $3. Of course if the better rolls zero 6s, which he does with probability 125/216, he loses $1. So the overall situation is

$$\frac{75}{216} \cdot 1 + \frac{15}{216} \cdot 2 + \frac{1}{216} \cdot 3 - \frac{125}{216} \cdot 1; t$$

that is, $17/216 = 7.87 cents, that is, a loss of about 7.87%. The same applies for the number 1, 2, 3, 4, and 5. So the shooter stands to lose 7.87%. This is a large percentage compared to the games we have previously discussed.

Blackjack

Example 1.14 The card games *blackjack, pontoon, and twenty-one* are for 3–10 players and are descended from the *Vingt-Un* game of the seventeenth century played in Spain. It spread to Germany, France, and Britain and became known as *pontoon*. When it spread to the USA in the nineteenth century, it became known as *blackjack*, and this is the most well-known of these games, especially in the context of casino games. Of course the rules and terminology used in blackjack games varies from casino to casino throughout the world. Our aim is definitely *not* to encourage you to play blackjack. Though of all casino games, blackjack probably offers the players *the chance to lose the least*; that is, the house advantage is less than it is for other

1.2 The Event Space and the Probability Space

games—usually between 0.5 and 1%. There are many books discussing blackjack including [4, 8, 17, 20, 29].

For our discussion, the game is played by the Dealer and two or more Players. The Players play against the Dealer, not against each other. Each Player is dealt two cards, face up and the Dealer is dealt one card. The game is played with one 52 card deck. (In most casinos, the game is played with 2, 4, 6, 8, or more decks.) Each card has a value. The cards $2, 3, \ldots, 10$ are assigned the value on their face, namely, $2, 3, \ldots, 10$. Jack, queen, and king have the value 10. The value of the ace is either 1 (called *hard*) or 11 (called *soft*), chosen by the Player (and can be changed by the Player during the game).

> For ease of our discussion, we shall assume the Dealer is female and the Players are male.

The aim of the game is to get the sum of the values of your cards as close to 21 as possible, but not strictly greater than 21, and to have the value greater than that of the Dealer. Players are permitted to draw additional cards to get closer to 21, but if they exceed 21 they lose whatever money they bet on that hand. When each Player has either exceeded 21 or wishes to be dealt no extra card, the Dealer is dealt a second card and if the value of the Dealer's hand is strictly less than 17, she will be dealt an extra card. An individual Player wins if he has cards of value 21 or has the value of his cards not strictly greater than 21 but strictly greater than that of the Dealer.

Once the cards have been dealt, the Player has four options:

1. *Stand.* The Player chooses not to receive any more cards.
2. *Hit.* The Player chooses to receive another card. The Player can choose to keep hitting until he stops or goes *Bust*; that is, goes over 21.
3. *Double.* The Player doubles his bet and gets dealt one more card.
4. *Split.* If the Player has two cards of the same value he can choose to split the cards, for example, if he has a 10 and a king. He doubles his stake (bet) as he has 2 hands and each hand receives a second card. (In the picture of the Blackjack table on the previous page, Player 1 has Split his hand as he had two cards of value 8.)

If the Dealer's card is an ace, there is an option to take *insurance*. This is in case the Dealer will have a Natural.

A *Natural* is an ace and a card of value 10 such as a 10 or jack, queen, or king (and thereby a total of 21). If a Player has a Natural, the Dealer immediately pays the player one and a half times the amount of his bet.

You might ask what is the probability of being dealt a Natural?

> We shall assume there is just one Player and one deck of cards. The number of aces is 4, and the number of cards of value 10 is 16. So the probability of a natural is $4/52 \cdot 16/51 + 16/52 \cdot 4/51 = 0.48\ldots$ (as the Ace or card of value

10 may be drawn first), that is, about 1/21. Of course the probability changes significantly if an ace, for example, has already been dealt.

So at each stage the Player has to decide whether he wishes to be hit (i.e. dealt an extra card) in order to get closer to 21 but not exceed 21. Our interest is in how a Player should decide whether to ask to be dealt an extra card. For this he uses probability.

(In practice, what card the Dealer has must not be ignored—for example, if it is a 2, and the Player has cards of value 12 or more, then Stand may be the best option.) But let us consider a simplification of the game, where the Player ignores the card that the Dealer has, so we get some feeling for what is going on. We shall consider the case that there is only *one Player plus the Dealer and that the casino is playing this game of blackjack with only one deck of 52 cards*.

First let us assume that the Player has been dealt two cards—a 6 and an 8—and so the total value of his hand is 14 and the Dealer's visible card is an 8. Now the Player must decide whether to be dealt another card. The Player knows that the Dealer will not stop until she reaches at least 17 and the Player knows his hand of 14 is not enough to beat 17. He decides to be dealt another card if there is a probability <0.5 that the value of his hand will go over 21. We know that if the next card dealt is an 8, 9, 10, jack, queen, or king, then the total value of his hand will be >21. In the whole deck there are 24 such cards. In his hand there is one card of this value. And the Dealer has an 8. So there are currently 22 such cards in the deck. The number of cards remaining in the deck is $52 - 3 = 49$. So the probability of the total value of his hand exceeding 21 is $\frac{22}{49}$ <0.5, and so he decides to ask to be dealt an extra card.

This time let us assume the Player has four aces, three 2s and one 4 for a total value of 14, as before, and the Dealer has an 8 as before. We know that if the next card dealt is an 8, 9, 10, jack, queen, or king, then the total value of his hand will be >21. In the whole deck there are 24 such cards. He has no such cards and the Dealer has one. So there remain 23 such cards in the deck. The number of cards remaining in the deck is $52 - 9 = 43$. So the probability of the total value of his hand exceeding 21 if dealt another card is $\frac{23}{43} > 0.5$ and so he decides not to ask to be dealt another card.

So we see from the above two cases that:
having the total value of his hand of 14 does not tell the Player whether to take another card or not.

Deciding whether you will go bust if you are dealt an extra card is not simply a question of counting the total value of your hand.

1.2 The Event Space and the Probability Space

> Now let us assume that the Player first two cards have a total value of 15 and the Dealer has any card. We shall consider what is the probability of the Player going bust if he is dealt another card.
>
> We know that if the next card dealt is a 7, 8, 9, 10, jack, queen, or king, then the total value of the Player's hand will be >21. There are 28 such cards in the single deck. As the Player's cards have total value 15, he can have at most one of these cards, and the Dealer might have one of these cards—so there at least 26 such cards in the deck.
>
> As the Player has at least 2 cards and the Dealer has one card, there are at most 49 cards still in the deck. So if the Player is dealt an extra card, the probability the total value of his hand exceeding 21 is $> \frac{26}{49} > 0.5$.
>
> So *if the Player's cards are of total value 15, and he is dealt one more card the probability of his busting is greater than 0.5. This is the case if the Player has cards of total value 16 or 17 or 18 or 19 or 20. HOWEVER, this calculation assumes that the player does not take Option 4 above and Split!*

> *We emphasize that we are focussing only on whether the Player taking an extra card has probability >0.5 of his going bust. But recall that the aim of the game is not only to avoid going bust, but to end up with a hand of value greater than that of the Dealer.*

In our previous discussion above, we were able to discover that if there is only one Player plus the Dealer, then the Player has a probability >0.5 of going bust if he asks for extra card when the total value of his hand is 15 or more. We also saw that if his hand has total value of 14, then whether he goes bust with an extra card depends on precisely which cards are in his hand. Having obtained the flavour of what results to expect, let us look at a slightly more realistic problem.

> This time we have 5 Players, one of whom is Ted. Each Player has been dealt 2 cards and Ted looks at the value of his cards and finds the total value is 16. Ted needs to decide if he will ask to be dealt an extra card. He knows that if the next card dealt to him is a 6, 7, 8, 9, 10, jack, queen, or king, then the total value of his hand will be >21. There are 32 such cards in the 52 card deck. He knows that each Player including himself could have a maximum of 2 such cards, that is 10 in all. The Dealer might have another one. So no matter what cards have actually been dealt, there must be at least 21 of these "big" cards remaining in the deck. Altogether 11 cards have been dealt, and so there are 41 cards remaining in the deck. So the probability that the extra card to be dealt will take the total value of his hand to >21 is $\frac{21}{41} > 0.5$. So if he is dealt an extra card, the probability of his going bust is >0.5.

We can tease a little more out of the above analysis. Let us assume that *each of the Players has been dealt at least two cards* and *Ted's cards have total value of 16 or 17 or 18 or 19 or 20*. Let us *assume* that *the total number of so-called "big" cards (that is 6, 7, 8, 9, 10, Jack, Queen, and King) that have been dealt to the Players and the Dealer is not more than 11* (as in the previous paragraph). So the number of "big" cards remaining in the deck is >21. The number of cards left in the deck <41. Then the probability that the extra card to be dealt will take the total value of Ted's hand to >21 is $>\frac{21}{41} > 0.5$. So *the probability that Ted will bust if he takes an extra card is >0.5.*

Optimum Strategy and Basic Strategy for Blackjack

In 1956 four mathematicians, Baldwin et al. [2] published what they called The Optimum Strategy in Blackjack. They addressed the specific problem of when to Stand and when to Hit. They did not address any other problems related to blackjack. This careful mathematical analysis opened the floodgates for others to come up with strategies for playing blackjack. If you want to see a basic strategy, see [4, p248]—such a strategy should allow you to lose your money slowly.

I made very clear at the beginning of this book that it is not my intention to provide strategies for winning at any of the games discussed in the book, including Blackjack.

The House Advantage

We have not made clear yet why the House (i.e. the casino) has an advantage. Consider what happens. The Dealer deals herself one card and the Player two cards. Next the Player decides whether to Stand, Hit, etc. At some point, the Player either Busts or Stands. If the Player Busts then the Dealer takes the Player's bet irrespective of whether the Dealer would have Bust with her second card. So the fact that the Player has the opportunity of busting first gives the Dealer the advantage! In a perfectly fair game, if the Player and Dealer both bust, then the Player would lose nothing, but this is not what happens. In essence this forces the Player to be extra careful not to Bust, thereby giving the Dealer a better chance of having a higher value of cards. For further discussion of this, see [29, p. 325]

1.2 The Event Space and the Probability Space

Billionaire Gambling

Bill Benter was regarded as a math whiz and was a billionaire gambler from Philadelphia, Pennsylvania. He tried his hand at blackjack and then in the 1980s turned to horse racing, particularly in Hong Kong. He developed a complex mathematical algorithm that could predict the probability of each horse in a race winning and using it he was very successful. His algorithm also told him the optimal amount to bet. He was so successful that the Jockey Club tried to ban him from betting, but he was able to fight them off in court. He retired from horse racing betting in 2001. The algorithm that Benter used did not guarantee that he won, but did mean that over time he won more than he lost. And he was rich enough not to worry about his losses, if in due course he would have profits. Few of us are in that category!

Horseracing Odds

Remark 1.3 Next we briefly explain horse racing *odds* [31].
Pictured below is the most famous horse in Australian history, the New Zealand born Australian trained *Phar Lap*, winning the most famous race in Australia, *The Melbourne Cup* in 1930. Also pictured is a rare triple dead heat in a 1952 harness race at *Freehold Raceway* in New Jersey, the oldest racetrack in the USA.

People place a bet on a particular horse in a particular race. Bets are often placed with a person called a *bookie* (short for bookmaker), a professional who facilitates gambling on most sporting events.
Let us look at the odds for a recent horse race:

Horse	Odds	Horse	Odds
1	51:1	6	11:5
2	81:1	7	101:1
3	17:2	8	51:1
4	22:5	9	23:5
5	22:5		

This says that if you place a bet of \$1 with the bookie on horse number 1 and it wins, then you receive \$51 plus your original bet of \$1, that is, \$52 in all. Similarly if you bet \$1 on horse number 5 and it wins, then you receive $\left(\frac{1}{5} \times 22\right) + 1 = \$5.40$.

How do odds relate to probabilities?

Converting odds to probabilities: if the odds are $a : b$, for some positive real numbers a and b, then the probability it corresponds to $\frac{1}{\frac{a}{b} + 1} = \frac{b}{a+b}$. So the probabilities for the above race are

Horse	Odds	Probability	Horse	Odds	Probability
1	51:1	$\frac{1}{52}$	6	11:5	$\frac{5}{16}$
2	81:1	$\frac{1}{82}$	7	101:1	$\frac{1}{102}$
3	17:2	$\frac{2}{19}$	8	51:1	$\frac{1}{52}$
4	22:5	$\frac{5}{27}$	9	23:5	$\frac{5}{28}$
5	22:5	$\frac{5}{27}$			

But we know the sum of all the probabilities should equal 1. Here, however, the sum is 1.027.... Why? This represents the built-in profit of the bookie of about 2.7%.

> Once again we have seen that if you gamble by placing a bet on a horse, over time you should expect to lose money. After all, you are providing the bookie's income!

We conclude this remark by showing that if the sum of the probabilities were strictly less than one, then it is possible to bet a certain amount on each horse and come out with a profit—which is why you will never find the sum of a bookie's probabilities less than one.

Let there be n horses in the race, and the odds are $a_1 : b_1, a_2 : b_2, \ldots, a_n : b_n$. So the corresponding probabilities are $p_1 = \frac{b_1}{a_1+b_1}, p_2 = \frac{b_2}{a_2+b_2}, \ldots, p_n = \frac{b_n}{a_n+b_n}$. Let us place a bet of x_1 on horse 1, x_2 on horse 2, ..., x_n on horse n, where we are yet to determine the best values of the x_i.

1.2 The Event Space and the Probability Space

The table below shows the Odds, Probabilities, Bets, and "Return"—the amount received if that horse wins plus the amount bet on that horse. If x_i is bet on horse i, and it wins then you receive $(x_i \times \frac{a_i}{b_i}) + x_i = x_i(\frac{a_i}{b_i} + 1) = x_i(\frac{a_i + b_i}{b_i})$. Let X be the total amount bet, that is, $X = x_1 + x_2 + \cdots + x_n$. Let G_i be the profit (or gain) on the bets on that race, that is, if horse i wins, then $G_i = \frac{x_i}{p_i} - X$. Define $P = p_1 + p_2 + \cdots + p_n$.

Horse	Odds	Probability	Bet	Return	
1	$a_1 : b_1$	$p_1 = \dfrac{b_1}{a_1 + b_1}$	x_1	$x_1 \left(\dfrac{a_1 + b_1}{b_1}\right) =$	$\dfrac{x_1}{p_1}$
2	$a_2 : b_2$	$p_2 = \dfrac{b_2}{a_2 + b_2}$	x_2	$x_2 \left(\dfrac{a_2 + b_2}{b_2}\right) =$	$\dfrac{x_2}{p_2}$
$\vdots$	$\vdots$	$\vdots$	$\vdots$	$\vdots$	
n	$a_n : b_n$	$p_n = \dfrac{b_n}{a_n + b_n}$	x_n	$x_n \left(\dfrac{a_n + b_n}{b_n}\right) =$	$\dfrac{x_n}{p_n}$

Now it is up to us to choose the values of $x_1, x_2, \ldots, x_n$. We shall choose $x_2, x_3, \ldots, x_n$ (but not x_1) such that

$$\frac{x_1}{p_1} = \frac{x_2}{p_2} = \cdots = \frac{x_n}{p_n}.$$

This means that $G_1 = G_2 = \cdots = G_n$. So we write $G = G_1 = G_2 = \cdots = G_n$. But we have $G = \frac{x_i}{p_i} - X$, and so $G + X = \frac{x_i}{p_i}$, which implies $p_i = \frac{x_i}{G + X}$. As $P = p_1 + p_2 + \cdots + p_n$ and $X = x_1 + x_2 + \cdots + x_n$, we have

$$P = \frac{x_1 + x_2 + \ldots x_n}{G + X} = \frac{X}{G + X}.$$

Thus $G = \frac{X}{P} - X = X\left(\frac{1}{P} - 1\right)$. Hence $G > 0$ if and only if $P < 1$.

This says that you can make a profit by betting appropriately on every horse if and only if $p_1 + p_2 + \cdots + p_n < 1$. And if the bets are chosen such that $\frac{x_1}{p_1} = \frac{x_2}{p_2} = \cdots = \frac{x_n}{p_n}$, then the profit is the same irrespective of which horse wins.

Macao

Macao, [21], is a card game dating back to at least the eighteenth century and apparently was popular with soldiers in the Austro-Hungarian Empire in the nineteenth century. It is regarded by some as a precursor of the popular casino game of *Baccarat*.

Each player is dealt a card by the banker and additional cards may be bought. The 2, 3, 4, 5, 6, 7, 8, 9 cards count as their face value, 10, jack, queen, and king count as 0, while ace counts as 1. The aim of the game is to get as close to nine points as possible.

If a player's first card is a 9, this is called a *grosser Schlag* and wins double unless the banker has the same, in which case the banker collects a double stake from each player except this player. Now 8 as a first card is called a *kleiner Schlag*.

Whoever goes bust is called *verkauft*, that is, ends up with more than nine points immediately loses their stake. If the banker goes bust, all players win. We shall not consider the game of Macao further as it became illegal.

We shall see that Macao has some similarity to Baccarat and some differences.

Baccarat

1897 illustration of Baccarat Players

The casino game of *Baccarat* appears in several James Bond books and movies.

There are three main varieties of Baccarat, namely, *Chemin de Fer* (a French version popular in Europe and Latin America), *Punto Banco* (developed in Havana in the 1940s and is the most popular variety in American casinos today), and *à Deux Tableux*. (Some stories suggest that while Napoleon was sacking Europe two forms of Baccarat were in vogue; one a banking game called Baccarat en Banque (Baccarat Deux Tableaux) the other Baccarat Chemin de Fer, a non-banking version. Chemin de Fer is a French phrase meaning railway, or literally iron path.)

1.2 The Event Space and the Probability Space

In Ian Fleming's 1958 novel Dr No, adapted for film by the British movie production company Eon Productions in 1962, Scottish-born actor Sean Connery as James Bond, is playing Baccarat Chemin de Fer.

There are many books describing the game of Baccarat, some including its history. We mention [4, 8, 17, 21].

For our purposes, Baccarat is a very simple game to understand. It is a card game played by two hands: the Player and the Banker. The game's objective is to predict which hand will win or whether there will be a tie.

As usual it is played with one or more standard packs of 52 cards. Up to eight packs is common. The bettors can bet on the Player or the Banker or a tie. Once the bets are placed, two cards are dealt face-up to both the Player and the Banker.

The cards 2, 3, 4, 5, 6, 7, 8, 9 have point value equal to their face number. 10s, jack, queen, and king have point value 0. Ace has point value 1.

The point value of the Player's hand and of the Banker's hand is the sum of the point values mod 10 of their respective hands. So if, for example, the Player has a 6 and a 5, then that hand has a point value of $(6 + 5 - 10 =)1$.

If the Player's hand or the Banker's hand has a point value of 8 or 9 then this is called a *natural* and no further cards are dealt.

If neither hand has a natural, a third card may be drawn according to predetermined rules. (To the newcomer these may seem complicated, but in practice they are straightforward or at least algorithmic.)

> The bettors do not need to remember these rules.

- The Player stands if the Player's hand has a point value of 6 or 7;
- the Player draws a card if the Player's hand has a point value of 0, 1, 2, 3, 4, or 5;
- the Banker stands if the Banker's hand has a point value of 7;
- if the Banker's hand has a point value of 0, 1, 2, 3, 4, 4, 5, or 6 the Banker may draw a third card according to the following circumstances:
- if the Player does not draw a third card, the Banker draws if their hand has a point value of 0, 1, 2, 3, 4, or 5 and stands if their hand has a point value of 6 or 7;
- if the Player draws a third card, then the Banker may draw a third card based on the following conditions:
- the Banker draws if the Banker's hand has a point value of 2 or less;
- if the Banker's hand has a point value of 3 and the player's third card is not an 8, the Banker draws;
- the Banker draws if the Banker's hand has a point value of 4 and the Player's third card is 2, 3, 4, 5, 6, or 7;
- the Banker draws if the Banker's hand has a point value of 5 and the Player's third card is 4, 5, 6, or 7;
- the Banker draws if the Banker's draws if the Banker's hand has a point value of 6 and the Player's third card is either 6 or 7.

After this has been completed, the hand with the highest point value wins.

If the Player's hand wins, the bettors who bet on the Player are paid out at odds of 1:1.

If the Banker's hand wins, then those who bet on the Banker pay a 5% commission to the House on their even money winnings.

If it is a tie, then those who bet on a tie are paid out at odds of 8:1.

> An analysis of the betting, shows that the Player bet has a House edge of 1.24%. The Banker bet has a House edge of 1.06%. The worst bet is the tie bet, which has a House edge of 14.4%, which is terrible.

Of course some casinos offer various other betting possibilities such as where the first cards dealt to the Banker are a pair, etc.

Ballot Box Problem

Next we present a rather surprising observation known as the *Ballot Box Problem*. The Ballot Box Problem was first published [32] in 1878 by William Allen Whitworth (1840–1905) who was an English mathematician and a priest in the Church of England. While an undergraduate, he became the founding editor in chief of the "Messenger of Mathematics" and continued as editor until 1880. His most important mathematical publication is the book *Choice and Chance: An elementary treatise on permutations, combinations and probability*, [33], (first published in 1867 and expanded in later editions). He introduced the now commonly used notation **E[X]** for the *expected value of a random variable X*

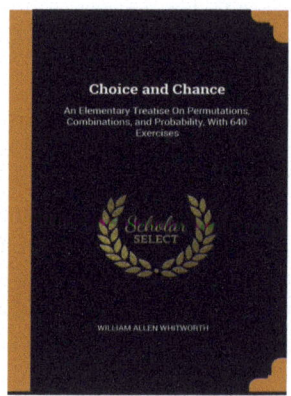

1.2 The Event Space and the Probability Space

The Ballot Box Problem was rediscovered in 1887 by Joseph Louis François Bertrand (1822–1900) who was a French mathematician who worked in differential geometry, economics, number theory, probability, and thermodynamics. The Ballot Box Problem is (unfortunately) referred to as Bertrand's ballot theorem. Bertrand was a Professor at the École Polytechnique and Collège de France. He was a member of the Paris Academy of Sciences and was its permanent secretary for 26 years. By the age of 17, he had two bachelor's degrees and a PhD. In 1907 he published the book [3].

Bertrand

Bertrand's paper outlined a proof using a recursion relation but remarked that it seems probable that it could be proved by a more direct method. Such a proof was given by Désiré André (1840–1917) who was a French mathematician and a student of Charles Hermite (1822–1901) and Bertrand. A variation of André's method is known as André's reflection method, although André did not use reflections.

Theorem 1.1 [Ballot Box Theorem] *In an election where candidate A receives p votes and candidate B receives q votes with $p > q$, the probability that A will be strictly ahead of B throughout the count is $\frac{p-q}{p+q}$.*

Proof. Observe firstly that for A to be strictly ahead of B throughout the count, there cannot be any point in the count where there is a tie.

Case 1: *The first vote goes to B.* Then the count must reach a tie at some point as A eventually wins. Observe that the probability that the sequence of votes begins with a B is $\frac{q}{p+q}$.

Case 2: *The first vote goes to A and at some point in the count reaches a tie.* Reflect the votes up to the tie, that is, change every A to B and every B to an A up to and including when the count reaches a tie. This reflection gives a new sequence which begins with a B and in due course reaches a tie. So there is a one-to-one correspondence between the sequences which begin with a B and the sequences which begin with an A and reach a tie. So the probability that the sequence of votes begins with an A and reaches a tie is also $\frac{q}{p+q}$.

Thus the probability that the count reaches a tie is $\frac{q}{p+q} + \frac{q}{p+q} = \frac{2q}{p+q}$.

But the probability that A always leads in the count equals 1− the probability that the counting reaches has a tie, that is, $1 - \frac{2q}{p+q} = \frac{p-q}{p+q}$. □

Remark 1.4 Let us consider for a moment what the Ballot Box Theorem tells us. If A wins strictly less than 75% of the vote, then there is a probability of greater than 0.5 that B will lead in the voting at some point in the count. In other words, if A does not receive strictly more than 3 times the number of votes that B receives, then there is a probability of greater than 0.5 that B will lead in the voting at some point in the count. Many, perhaps most, people would be surprised by this.

[In the 1969 Australian Federal election at one point the count showed that the Coalition had won 47 seats and the Labor party had won 62 seats—a huge margin. In fact, at the end of the counting, the Coalition won the election.]

Weather Forecasting and Chaos Theory

Today when I read the weather forecast for tomorrow no doubt it will predict the temperature range and a *probability* of rain. What exactly is meant by this probability? When it says there is an 80% chance of rain in my city, does it mean that 80% of my city will experience rain and the other 20% of the city will not? Of course it does not mean this! When did forecasting of the weather begin to include probabilities? It is difficult to be precise, but certainly a key figure was the Australian astronomer William Ernest Cooke (1863–1947). (See [7].)

Cooke

To understand today's forecasting we need to look briefly into *chaos theory*. (It would be folly to try to give a definitive history of chaos, a term used in the book of *Genesis* in the *Bible* and *Hun-Tun* (translated as chaos) in *Taoism* ([13]), a philosophical tradition dating back 2,200 years in China to the *Han Dynasty*. Here we focus on the twentieth century. The interested reader may care to read my introductory chaos theory material in [18, Appendix 3].)

Jules Henri Poincaré (1854–1912), one of France's greatest mathematicians, is acknowledged as one of the founders of a number of fields of mathematics including modern nonlinear dynamics, ergodic theory, and topology. His work laid the foundations for chaos theory. He stated in his 1903 book, [23], *If we knew exactly the laws of nature and the situation of the universe at its initial moment, we could predict exactly the situation of that same universe at a succeeding moment.*

Poincaré

1.2 The Event Space and the Probability Space

But even if it were the case that the natural laws had no longer any secret for us, we could still only know the initial situation approximately. If that enabled us to predict the succeeding situation with the same approximation, that is all we require, and we should say that the phenomenon had been predicted, that it is governed by laws. But it is not always so; it may happen that small differences in the initial conditions produce very great ones in the final phenomena. A small error in the former will produce an enormous error in the latter. Prediction becomes impossible. What Poincaré described quite precisely has subsequently become known colloquially as the *butterfly effect*, an essential feature of chaos.

In 1952 Collier's magazine published a short story called "A Sound of Thunder" by the renowned author, Ray Bradbury (1920–2012). In the story, http://www.lasalle.edu/~didio/courses/hon462/hon462_assets/sound_of_thunder.htm
a party of rich businessmen use time travel to journey back to a prehistoric era and go on a safari to hunt dinosaurs. However, one of the hunters accidentally kills a prehistoric butterfly, and this innocuous event dramatically changes the future world that they left. This was perhaps the incentive for a meteorologist's presentation in 1973 to the American Association for the Advancement of Science in Washington, D.C. being given the name "Predictability: Does the flap of a butterfly's wings in Brazil set off a tornado in Texas?"

Bradbury

On April 16, 2008, MIT News published "Edward Lorenz, father of chaos theory and butterfly effect, dies at 90". I quote from that announcement: "Edward Lorenz, an MIT meteorologist who tried to explain why it is so hard to make good weather forecasts and wound up unleashing a scientific revolution called chaos theory, died April 16 of cancer at his home in Cambridge. He was 90.

A professor at MIT, Lorenz was the first to recognize what is now called chaotic behaviour in the mathematical modelling of weather systems. In the early 1960s, Lorenz realized that small differences in a dynamic system such as the atmosphere–or a model of the atmosphere–could trigger vast and often unsuspected results.

These observations ultimately led him to formulate what became known as the butterfly effect–a term that grew out of an academic paper he presented in 1972 entitled: 'Predictability: Does the Flap of a Butterfly's Wings in Brazil Set Off a Tornado in Texas?'

Lorenz's early insights marked the beginning of a new field of study that impacted not just the field of mathematics but virtually every branch of science–biological, physical and social. In meteorology, it led to the conclusion that it may be fundamentally impossible to predict weather beyond two or three weeks with a reasonable degree of accuracy.

Some scientists have since asserted that the twentieth century will be remembered for three scientific revolutions–relativity, quantum mechanics and chaos."

Edward Norton Lorenz (1917–2008) discovered sensitivity to initial conditions by accident. He was running on a computer a mathematical model to predict the weather. Having run a particular sequence, he decided to replicate it. He re-entered the number from his printout, taken part-way through the sequence, and let it run. What he found was that the new results were radically different from his first results. Because his printout rounded to three decimal places, he had entered the number 0.506 rather than the six digit number 0.506127. Even so, he would have expected that the resulting sequence would differ only slightly from the original run. Since repeated experimentation proved otherwise, Lorenz concluded that *the slightest difference in initial conditions made a dramatic difference to the outcome*. So prediction was in fact impossible. Sensitivity to initial conditions, or the butterfly effect, had been demonstrated to be not just of theoretical importance but in fact of practical importance in meteorology. It was a serious limitation to predicting the weather—at least with that model.

So we see that while we can measure the weather conditions all over the world approximately, we cannot measure the precise values. And we now know that even the tiniest amount of approximation can impact the prediction dramatically. This issue cannot be made to go away by better and better accuracy of the measurements. So what is to be done? The answer is what has become known as *ensemble forecasting*. Ensemble forecasting is a form of *Monte Carlo analysis*. (Monte Carlo methods are a class of computational algorithms that rely on repeated random sampling to obtain numerical results.) The notion is to use an averaging of a number of simulations to address the sensitivity to initial conditions.

Let me use the description provided by "The Met Office", which is the national meteorological service for the United Kingdom. "To forecast the weather we first gather observations from around the world to measure what the atmosphere is doing. We use these observations to set up a computer simulation of the atmosphere that represents what is happening right now. The model then calculates how the atmosphere will evolve over the coming days. Unfortunately, due to chaos, small unknowns in our observed atmosphere can grow rapidly to give large uncertainties in the forecast.

Over the last 15 years, the Met Office has developed sophisticated techniques to understand these uncertainties, called ensemble forecasts. This means we run our simulations many times instead of just once, from very slightly different starting conditions. The range of different outcomes gives us a measure of how confident or uncertain we should be in the overall forecast. On some occasions the uncertainty is quite small and we can be confident—other times much less so. This can help decision makers manage the risks associated with the weather."

Typhoon Haiyan, 2013

To oversimplify this dramatically, we might say if 8 of 10 simulations indicate rainfall of over 1cm, the chance of rain exceeding 1cm could be estimated to be 80%. Of course the Met Office uses analysis that is more sophisticated than this.

Simpson's Paradox

This paradox is often used in the teaching of mathematical statistics to illustrate the care we need to take when interpreting data and to avoid jumping to conclusions.

Edward H. Simpson (1922–2019) described this phenomenon in a paper in 1951 but the statisticians Karl Pearson (1857–1936), Alice Lee (1858–1939), and Lesley Bramley-Moore (1831–1918) in 1899 and George Udny Yule (1871–1951) in 1903 had already mentioned similar effects.

On June 10, 1898, Alice Lee marched into the all-male Anatomical Society meeting at Trinity College in Dublin and pulled out a measuring instrument. She then began to measure the size of all 35 consenting society members' heads. She ranked their skulls from largest to smallest to discover that some of the most well-regarded intellects turned out to possess rather small skulls. In due course Alice Lee became a PhD student of the esteemed statistician Karl Pearson. In her PhD dissertation she did a study of male and female intellectual difference. Until then experts had asserted that intellectual capability was related to cranial size. As a result, women had inferior intellectual capability. Lee provided the most sophisticated criticism of cranium science to date. Within a decade of publishing her findings in 1900, the field of *craniology*—and with it, the days of measuring skulls to interpret biological human difference—would be no more. This applied both to sex and race.

George Udny Yule was a British statistician, particularly known for the *Yule–Simon distribution* (or Yule distribution) is a discrete probability distribution named after Udny Yule and Herbert A. Simon. Simon originally called it the Yule distribution.

The name Simpson's paradox was introduced, in [6], by the Canadian statistician Colin Ross Blyth (1922–2019) in 1972.

In probability and statistics, the Yule–Simon distribution is a discrete probability distribution named after Udny Yule and Herbert A. Simon. Simon originally called it the Yule distribution. Herbert Alexander Simon (1916–2001) was an American economist and political scientist whose main research interest was in decision-making. He received the Nobel Prize in Economics in 1978.

Edward Hugh Simpson was introduced to mathematical statistics as a cryptanalyst at Bletchley Park (1942–45). He wrote the paper "The Interpretation of Interaction in Contingency Tables" while a postgraduate student at the University of Cambridge in 1946 with Maurice Stevenson Bartlett (1910–2002) as his tutor and published it in the *Journal of the Royal Statistical Society* in 1951 at Bartlett's request because Bartlett wanted to refer to it. Maurice Stevenson Bartlett (1910–2002) was an

English statistician who is known for work in statistical inference and multivariate analysis. Simpson's paper considered what is now known as the *Yule-Simpson effect* or *Simpson's paradox* or the *amalgamation paradox* or the *reversal paradox*.

Blyth [6] demonstrated the paradox in the following way.

There is a new treatment N and a standard treatment S for a serious complaint. Patients are in two groups A and B, according to their residence is local or in Chicago. If the patient survives with a treatment, they are said to alive A and if not they are said to be dead D. The table below is a record of treatment results.

	Chicago patients		Local patients		Total patients	
	Standard	New	Standard	New	Standard	New
Dead	950	9,000	5,000	5	5,950	9,005
Alive	50(5%)	1,000(10%)	5,000(50%)	95(95%)	5,050(46%)	1,095(11%)

(i) For Chicago patients, 5% of the patients survive (alive) with the standard treatment, while only 10% of patients survive with the new treatment. So *the new treatment is clearly better for the Chicago patients*.

(ii) For the Local patients, 50% of the patients survive (alive) with the standard treatment, while 95% of patients survive with the new treatment. So we see *the new treatment is clearly better for the local patients*.

But surprisingly, *for the total patient population, 46% survive with the standard treatment, while only 11% survive with the new treatment. So the standard treatment is very much better.*

> In the literature there are many such examples. The moral is as follows: as the Hollywood writer, director Jerry Belson (1938–2006) observed *Never ASSUME, because when you ASSUME, you make an ASS of U and ME.*

Note that if the number of patients in each of the two groups—local and Chicago—had been equal, then the paradox would not have occurred. However, this is not really a solution as in "real life" the numbers may not be equal. But we now notice that the number in each group is a variable, the impact of which we did not take account. This variable is called an *extraneous variable* and here is, in fact a *confounding variable* in that it clearly impacts on the result of the experiment.

Statisticians need to be aware of this paradox and design their experiments and analyses with great care so as not to obtain spurious results. This is not the place to say more about this very important topic.

Problems

Klondike Solitaire

1.12 Solitaire, also known as *Patience*, is a game played usually by one player. There are many varieties of the game. The most common variety is *Klondike Solitaire*. The rules can be found at https://en.m.wikipedia.org/wiki/Klondike_(Solitaire) and a video explaining the game can be watched on https://youtu.be/gHZc5O-e0dA

This game uses all 52 cards of the standard pack. The cards in ascending order are ace, 2, 3, 4, 5, 6, 7, 8, 9, 10, jack, queen, and king and each black card (spades or clubs) in this list can be placed on the red card (diamonds or hearts) next on this list and vice versa for red on black. Initially the cards are laid out in seven columns with the first column having one card, the second two cards, ..., the 7th column has seven cards.

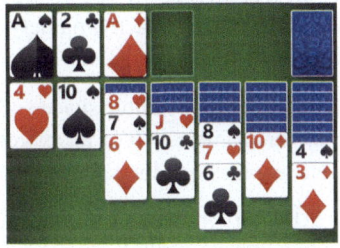

At the beginning only the top card in each column is face up. There are four foundation piles, one for each of the suits: spades, clubs, diamonds, and hearts. During the game, these foundation piles are filled, each starting with an ace, then a 2, ..., king. At the start there are $1 + 2 + \cdots + 7 = 28$ cards used in the columns, leaving 24 cards remaining in the stack. The picture above shows the situation when the game has been played for a few turns. Next the third card from the top of stack is turned face up to see if it can be put on one of the columns or on a foundation pile (e.g. a 3 of spades would be placed on the 4 of hearts; an ace of diamonds would be placed on the empty foundation pile; a 3 of clubs would be placed on the clubs foundation pile). If it cannot, then the card is considered not useful. Is the probability that this card is useful at this point in the game above greater than $\frac{1}{3}$?

1.13 Gaucher's disease (GD) is another example of an autosomal recessive inheritance disease of the type described in Example 1.13. The disease is caused by a recessive mutation in the GBA gene located on chromosome 1 and is named after the French physician Philippe Charles Ernest Gaucher (1854–1918). About 1% of people in the USA are carriers. It is much more common amongst Ashkenazi Jews. If a man who has Gaucher's disease has a child with a woman who does not have Gaucher's disease and is also not a carrier, what is the probability of each of the following: (a) the child has Gaucher's disease; (b) the child is a carrier of Gaucher's disease; (c) the child does not have Gaucher's disease and is not a carrier?

1.14 Darlene is calculating the probability she has Gaucher's disease (see previous problem) or is a carrier. She knows that none of her four grandparents have Gaucher's disease and that precisely n of them are a carrier of one Gaucher's disease gene, where $n \in \{0, 1, 2, 3, 4\}$. Calculate the probability that she (a) has Gaucher's disease or (b) is a carrier of a Gaucher's disease gene if (i) $n = 0$, (ii) $n = 1$, (iii) $n = 2$, (iv) $n = 3$, (v) $n = 4$.

1.15 Prove Proposition 1.4.

1.16 I have two fair coins. If I throw both coins in the air and then observe after they land the side that is facing up on both coins. What is the probability that the result will be (a) both coins land heads up; (b) both coins land tails up; (c) both coins land with the same side up?

1.17 I throw two fair dice in the air and after they land observe the number of dots on the side facing up on each die. What is the probability that the sum of these two numbers is a prime number? (Recall the definition that a positive integer $m \in \{1, 2, \ldots, n, \ldots\}$ is a *prime number* if there are precisely two positive integers which divide m exactly, namely, 1 and m. So $2, 3, 5, 7, 11, \ldots$ are prime numbers while $1, 4, 6, 8, 9, 10, 12, \ldots$ are not prime numbers.)

1.18 Let Ω be a sample space and Σ be an event space on Ω. Let $A, B, C \in \Sigma$. Noting that, for example, AB' is an expression of the fact that A occurs but B does not occur, write expressions for each of the following:

(i) A, B, and C occur;
(ii) only B occurs;
(iii) A and C occur but B does not occur;
(iv) precisely two of A, B, and C occur;
(v) at least one of A, B, and C occur;
(vi) none of A, B, C occurs.

1.19 Let Σ be an event space. Prove the following.

(i) if $A, B, C \in \Sigma$, then

$$P(A \cup B \cup C) = P(A) + P(B) + P(C) - P(AB) - P(AC) - P(BC) + P(ABC).$$

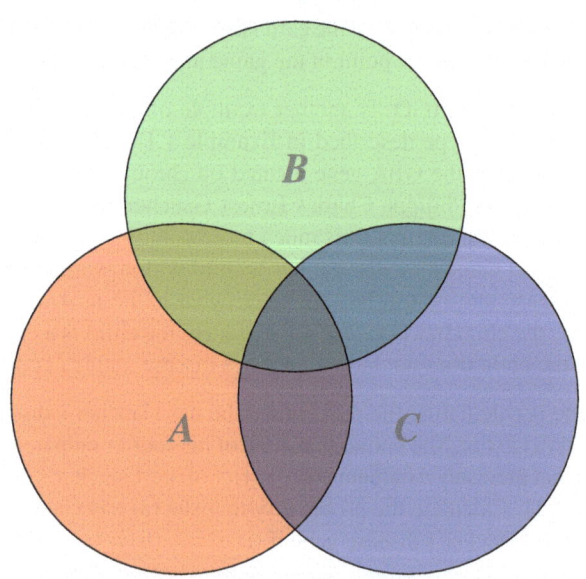

1.2 The Event Space and the Probability Space

(ii) If $A_1, A_2, \ldots, A_n \in \Sigma$, then

$$P\left(\bigcup_{i=1}^{n} A_i\right) = S_1 - S_2 + S_3 - \cdots + (-1)^{n+1} S_n, \text{ where}$$

$$S_1 = \sum_{i=1}^{n} P(A_i), \quad S_2 = \sum_{i<j \le n} P(A_i \cap A_j),$$

$$S_3 = \sum_{i<j<k \le n} P(A_i \cap A_j \cap A_k), \ldots, S_n = P\left(\bigcap_{i=1}^{n} A_i\right).$$

Note that $P(\bigcup_{i=1}^{n} A_i)$ is the probability that *at least one* of the events $A_1, A_2, \ldots, A_n$ occur.

1.20 In the game of French roulette as described in Example 1.12, we saw that there is a variety of common betting options. In these options the player generally bets not on one number but rather a set of numbers.

(i) We consider a simple example where the player bets on $n \in \mathbb{N}$ numbers on the wheel. The payout for a win on a bet of \$1 on n numbers is \$$\frac{36}{n}$. What is the probability of this being a winning bet and what is the expected loss?

(ii) Often the player makes a combination bet. For example, they may bet \$$a_1$ on a certain set of n_1 numbers, \$$a_2$ on another set of n_2 numbers, $\ldots$, and \$$a_k$ on a set of n_k numbers, for some $a_1, a_2, \ldots, a_k \in \mathbb{N}$ and $n_1, n_2, \ldots, n_k \in \mathbb{N}$. The payout for such a combination bet is \$$\sum_{i=1}^{k} \frac{36 a_i}{n_i}$. What is the probability of this being a winning bet and what is the expected loss?

1.21 Again in the game of French roulette as discussed in the previous problem, we shall consider combination bets.

(i) One combination bet is called a Zero bet, as the player bets on the numbers close to 0 on the wheel, namely, the set $A = \{12, 35, 3, 26, 0, 32, 15\}$ of 7 numbers. The player bets a total of \$4 as follows: \$1 on the pair $\{0, 3\}$, \$1 on the pair $\{12, 15\}$, \$1 on the pair $\{32, 35\}$ and \$1, on the number $\{26\}$. Using (i) and (ii) of the previous problem, show that the payout if the ball stops in the set A of 7 numbers is \$90, and calculate the probability of this being a winning bet and the expected loss?

(ii) Another combination bet is the Voisins du Zero bet. Here the player bets on the 17 numbers lying on the wheel between 22 and 25, including 22 and 25. This is the set $B = \{22, 18, 29, 7, 28, 12, 35, 3, 26, 0, 32, 15, 19, 4, 21, 2, 25\}$. The player bets a total of \$9 as follows: \$2 on the three numbers $\{0, 2, 3\}$, \$1 on the pair $\{4, 7\}$, \$1 on the pair $\{12, 15\}$, \$1 on the pair $\{18, 21\}$, \$1 on the pair $\{19, 22\}$, \$2 on the $\{25, 26, 28, 29\}$ corner, and \$1 on the pair $\{32, 35\}$. Using (i) (ii) of the previous problem, show that the payout for a winning Voisins bet is \$132, and calculate the probability of this being a winning bet and the expected loss.

1.22 Consider a horse race with the odds as indicated in the table below. Verify that it is possible to bet on every horse in the race in such a way that you make a profit irrespective of which horse wins. Further, verify that your profit will be in excess of 20% of the total amount bet on all the horses.

Horse	Odds	Horse	Odds
1	60:1	6	6:1
2	90:1	7	100:1
3	10:1	8	50:1
4	5:1	9	7:1
5	3:1		

1.23 In the game of Baccarat, the Player stands if they have a hand with point value of 6 or 7. Assuming they are playing with one deck of cards, how many different hands have a point value of 6 or 7. Deduce from this the probability of the Player having a hand of point value 6 or 7.

1.3 Credit for Images

- Ars Conjectandi. Public Domain.
- Autosomal recessive. Creative Commons Attribution-Share Alike 4.0 International license. https://commons.wikimedia.org/wiki/File:Autosomal_dominant_and_recessive.svg. File modified only so that it includes only the recessive case.
- Ernest William Barnes. Public Domain.
- Martinus Willem Beijerinck. This file is licensed under the Creative Commons Attribution 4.0 International license.
  https://commons.wikimedia.org/wiki/File:Martinus_Willem_Beijerinck.png
- Daniel Bernoulli. Public Domain.
- Johann Bernoulli. Public Domain.
- Sergei Bernstein. Public Domain.
- Joseph Louis François Bertrand. Public Domain.
- Irénée-Jules Bienaymé. Public Domain.
- Blackjack. Public Domain.
- Félix Édouard Justin Émile Borel. Public Domain.
- Ray Bradbury. "'All I ask is that you put 'Photo by Alan Light' if you can. That's not essential, but appreciated."
  https://en.wikipedia.org/wiki/Ray_Bradbury#/media/File:Ray_Bradbury_(1975)_-cropped-.jpg
- Calcul Des Probabilités. Fair Use.
  https://www.amazon.com.au/s?k=Calcul+Des+Probabilites&i=stripbooks&ref=nb_sb_noss
- Girolamo Cardano. Public Domain.
- Charles Chamberland. Public Domain

1.3 Credit for Images 51

- Jean Baptiste-Siméon Chardin. Public Domain.
- William Ernest Cooke. Public Domain.
- Francis Harry Compton Crick. This file was published in a Public Library of Science journal. Their website states that the content of all PLOS journals is published under the Creative Commons Attribution 4.0 license (or its previous version depending on the publication date), unless indicated otherwise.
https://commons.wikimedia.org/wiki/File:Francis_Crick_crop
- Abraham De Moivre. Public Domain.
- Augustus De Morgan. Public Domain.
- Charles John Huffam Dickens. Public Domain.
- DNA $\longrightarrow$ DNA. his file is licensed under the Creative Commons Attribution-Share Alike 3.0 Unported license.
https://commons.wikimedia.org/wiki/File:Eukaryote_DNA-en.svg
- Double Helix: Double stranded DNA with coloured bases. This file is licensed under the Creative Commons Attribution-Share Alike 4.0 International license.
https://commons.wikimedia.org/wiki/File:Double_stranded_DNA_with _coloured_bases.png
https://commons.wikimedia.org/wiki/File:Dreidel_001.jpg
- Ancient Russian Enamel Zolotnik Dreidel. Creative Commons Attribution-Share Alike 4.0 International license
https://commons.wikimedia.org/wiki/File:Russian_Dreidel_(1).jpg
- Paul Erdös playing Go, November 1979. Copyright held by Sidney A. Morris.
- Euchered lithograph 1884 from the Library of Congress. Public Domain.
- Sir Ronald Aylmer Fisher. Public Domain.
- Fragment of Papyrus from Euclid's Elements. Public Domain.
- French Roulette Wheel. Creative Commons Attribution-Share Alike 4.0 International License. https://commons.wikimedia.org/wiki/File:Roulette_casino.jpg
- Rosalind Elsie Franklin. This file is licensed under the Creative Commons Attribution-Share Alike 4.0 International license.
https://commons.wikimedia.org/wiki/File:Rosalind_Franklin.jpg
- Martin Gardner. This file is licensed under the Creative Commons Attribution-Share Alike 2.0 Germany license.
https://en.wikipedia.org/wiki/Martin_Gardner#/media/File:Martin_Gardner. jpeg
- Johann Carl Friedrich Gauss. Public Domain.
- Ralph William Gosper Jr. Creative Commons Attribution 2.0 Generic license.
https://commons.wikimedia.org/wiki/File:Bill_Gosper_2006.jpg
- Friedrich Robert Helmert. Public Domain.
- David Hilbert. Public Domain.
- Cristiaan Huygens. Public Domain.
- Influenza virus. Public Domain.
- Dmitri Iosifovich Ivanovsky. Public Domain. This work is in the public domain in Russia according to article 1281 of Book IV of the Civil Code of the Russian Federation No. 230-FZ of December 18, 2006, articles 5 and 6 of Law No. 231-FZ of the Russian Federation of December 18, 2006 (the Implementation

Act for Book IV of the Civil Code of the Russian Federation).
https://en.wikipedia.org/wiki/File:Ivanovsky.jpg
- Andrej Nikolajewitsch Kolmogorov. Copyright MFO.
https://opc.mfo.de/detail?photoID=7493
- Pierre-Simon Laplace. Public Domain.
- Johannes Friedrich Miescher. Public Domain
- de Montmort's book. Public Domain
- Kary Banks Mullis. Public Domain.
https://commons.wikimedia.org/wiki/File:Kary_Mullis.jpg
- Nicole Oresme. Public Domain.
- Blaise Pascal. Permission is granted to copy, distribute and/or modify this document under the terms of the GNU Free Documentation License.
https://commons.wikimedia.org/wiki/File:Blaise_Pascal_Versailles.JPG
- Louis Pasteur. Public Domain.
- Karl Pearson. Public Domain. Public Domain.
- Phar Lap. Public Domain.
- Edgar Allan Poe. Public Domain.
- George Pólya. This file is licensed under the Creative Commons Attribution 2.0 Generic license.
https://commons.wikimedia.org/wiki/File:George_P%C3%B3lya_ca_1973.jpg
- Jules Henri Poincaré. Public Domain.
- Principia Mathematica. Photograph, Andrew Dunn, 5 November 2004. http://www.andrewdunnphoto.com/ Creative Commons Attribution-Share Alike 2.0 Generic license.
- Typhoon Haiyan approaching the Philippines on November 7, 2013. Public Domain
https://en.wikipedia.org/wiki/Storm#/media/File:Haiyan_2013-11-07_ 0420Z.jpg
- Richard Edler von Mises. This work is free and may be used by anyone for any purpose. Konrad Jacobs, Erlangen—https://opc.mfo.de/detail?photo_id=2896
- James Dewey Watson. Public Domain.
- Women's Whist Club Congress Drawing. Public Domain.
- $\dfrac{1}{\zeta(x)}$ Graph. Graph produced using WolframAlpha.
- Sic Bo Table. Public Domain.
- Lizzie Maggie. Public Domain.
- The Landlord's Game. This file is licensed under the Creative Commons Attribution 2.5 Generic license. https://commons.wikimedia.org/wiki/File:Landlords_Game_board_based_on_1924_patent.png
- 1897 Illustration of Baccarat. Public Domain.
- Craps Table Layout. Permission is granted to copy, distribute and/or modify this document under the terms of the GNU Free Documentation License, Version 1.2 or any later version published by the Free Software Foundation; with no Invariant Sections, no Front-Cover Texts, and no Back-Cover Texts. A copy of the license is included in the section entitled GNU Free Documentation License.
https://commons.wikimedia.org/wiki/File:Craps.svg

References

1. Athreya, K.B., Lahiri, S.N.: Measure Theory and Probability Theory. Springer, New York (2006)
2. Baldwin, R.R, Cantey, W.E., Maisel, H., McDermott, J.P.: The optimum strategy in blackjack. J. Am. Statist. Assoc. **51**(275), 429–439 (1956)
3. Bertrand, J.: Solution d'un problème. Comptes Rendus de l'Acad 'emie Sciences de Paris !**05**, 369 (1887)
4. Bollman, M.: Mathematics of the Big Four Casino Table Games. CRC Games, Boca Raton, (2022)
5. Billingsley, P.: Probability and Measure, 3rd edn. Wiley, New York (1995)
6. Blyth, C.R.: On Simpson's paradox and the sure-thing principle. J. Amer. Statistical Assoc. **67**(338), 364–366 (1972)
7. Cooke, W.E.: Weighting forecasts. Monthly Weather Rev. **34**, 274–275 (1906)
8. Crawford, D.: A Beginners Gambling Guide at the Casino: Learn how to play Blackjack, Craps, Roulette & Baccarat. Derrick Crawford (2023)
9. Doetsch, G.: Introduction to the Theory and Application of the Laplace Transformation. Springer, Berlin (1974)
10. Dummett, M., McLeod, J.: A History of Games Played with the Tarot Pack. The Edwin Mellen Press, Lewiston (2004)
11. Efron, B.: R.A. Fisher in the 21st century. Statist. Sci. **13**(2), 95–114 (1998)
12. Epstein, R.A.: The Theory of Gambling and Statistical Logic, 2nd edn. Academic, Amsterdam (2009)
13. Girardot, N.J.: Myth and Meaning in Early Taoism: The Theme of Chaos (hun-tun). University of California Press, Berkeley (1983)
14. Hald, A.: A History of Mathematical Statistics from 1750 to 1930. Wiley, New York (1988)
15. Hald, A.: A History of Probability and Statistics and their Applications Before 1750. Wiley, New Jersey (2003)
16. Halmos, P.: Naive Set Theory. Van Nostrand Reinhold Company, New York (1960)
17. Mendelson, P.: The Mammoth Book of Casino Games. Constable & Robbins, London (2010)
18. Morris, S.A.: Topology Without Tears (1985–2023). www.topologywithouttears.net
19. Morris, S.A.: Hilbert 13: are there any genuine continuous multivariate real-valued functions? Bull. Am. Math. Soc. **58**(1), 107–118 (2021). https://doi:10.1090/bull/1698
20. Packel, E.: The Mathematics of Games and Gambling. MAA Press, Nelson (2006)
21. Parlett, D.: The Penguin Book of Card Games. Penguib Books, London (2008)
22. Parlett, D.: Oxford History of Board Games. Echo Point Books & Media, Brattleboro, Reprint: Revised edition (2018)
23. Poincarè, H.: Science and Method; Translated and Republished. DoverPress, New York (2003)

24. Pap, E. (ed): Handbook of Measure Theory. North Holland, Amsterdam (2002)
25. Pablo, E.M.: The Fibonacci Strategy: How to Make Money Playing Roulette. Independently Published by E.M. Pablo (2019)
26. Posamentier, A.S., Lehmann, I.: The Fabulous Fibonacci Numbers. Prometheus, Buffalo (2023)
27. Rigal, B.: Card Games for Dummes, 2nd edn. with Foreword by O. Sharif. Wiley, Hoboken (2005)
28. Rowntree, D.:Probability Without Tears. Scribner's, New York (1984)
29. Scarne, J.: Scarne's Complete Guide to Gambling. Simon and Schuster, New York (1961)
30. Scarne, J.: Scarne on Dice, Eighth Revised Edition. Stackpole Books, Mechanicsburg (1974)
31. Sohail, S: The math behind betting odds and gambling (2022). https://www.investopedia.com/articles/dictionary/042215/understand-math-behind-betting-odds-gambling.asp
32. Whitworth, W.A.: Arrangements of m things of one sort and n things of another sort under certain conditions of priority. Messenger of Math. **8**, 105–114 (1878)
33. Whitworth, W.A.: Choice and Chance. Literary Licensing LLC, Whitefish (2014). New Release of Book Edition of 1870

Chapter 2
Permutations and Combinations

Abstract

In this second chapter, we introduce permutations and combinations and selection with and without replacement. Then we examine the games of bridge, poker, cribbage, euchre, rummy, dreidels, and the Australian game two-up. We emphasize the importance of bluffing in poker and the central role of bidding in bridge. A member of the rummy family is the game mahjong, which is played today by hundreds of millions of women, men, and children throughout the world. We spend quite some time making it easy for the reader with no knowledge of Chinese to understand the game. American mahjong, known as mah jongg, is introduced, and its role in modern American culture is referred to. Voltaire discovers a serious flaw in French lotto and becomes a millionaire. We mention the famous Dead Man's Poker Hand of Wild Bill Hickok; the £1,000 challenge to all takers in the game of Whist of the Earl of Yarborough in the 1800s (equivalent today to more than US$100,000); the seventeenth century Dice Problem of Samuel Pepys answered by his Royal Society colleague Sir Isaac Newton; the well-known, but surprising, birthday paradox; and even combination locks, lotto, powerball, bingo, and ChatGPT summarizing the rules of cribbage.

2.1 Permutations and Combinations

You probably have learned about permutations and combinations at high school. But do not be afraid if you cannot recall everything you learnt as I shall cover it afresh. In this section there are only a few things to learn and be sure you know:

  (i) what is a permutation;
 (ii) what is a combination;
(iii) the difference between permutations and combinations;
(iv) selection/sampling with replacement;
 (v) selection/sampling without replacement;
(vi) how to evaluate the permutation P_r^n;
(vii) how to evaluate the combination $\binom{n}{r}$;

(viii) the relationship between P^n_r and $\binom{n}{r}$.

Definition 2.1 Let S be any finite or infinite set. A *permutation* of the set S is a one-to-one mapping of S onto itself.

Example 2.1 Consider the following examples:

(i) Let $f : \mathbb{N} \to \mathbb{N}$ be given by $f(x) = 2x$, for all $x \in \mathbb{N}$. Then f is a one-to-one function, but is not onto, since there is no $x \in \mathbb{N}$ such that $f(x) = 1 \in \mathbb{N}$. So f is not a permutation of $\mathbb{N}$.
(ii) Let $f : \mathbb{R} \to \mathbb{R}$ be given by $f(x) = |x|$, for all $x \in \mathbb{R}$. Then f is not onto and is not one-to-one. So f is not a permutation of $\mathbb{R}$.
(iii) Let $f : \mathbb{R} \to \mathbb{R}$ be given by $f(x) = -x$, for all $x \in \mathbb{R}$. Then f is both one-to-one and onto. So f is a permutation of $\mathbb{R}$.
(iv) Let $f : \{1, 2, \ldots, 10\} \to \{1, 2, \ldots, 10\}$ be given by $f(1) = 10$, $f(2) = 9, \ldots, f(10) = 1$; that is, $f(n) = 11 - n$, for $n \in \{1, 2, \ldots, 10\}$. Clearly f is one-to-one and onto, and so is a permutation.
(v) Let i be a fixed number in $\mathbb{N}$ and $S = \{a_1, a_2, \ldots, a_n, \ldots\}$. Further let f be the mapping of S into itself given by $f(a_i) = a_{i+1}$, $f(a_{i+1}) = a_i$ and $f(x) = x$, otherwise. Then f is a permutation of S and is called a *two-cycle*, and written $(i, i + 1)$. More generally, if $k \in \mathbb{N}$ and f maps a_i to a_{i+1}, a_{i+1} to $a_{i+2}, \ldots a_{i+k-1}$ to a_{i+k}, a_{i+k} to a_i, and each other x in S to x, then f is a permutation of S called a *k-cycle* and is written $(i\,(i + 1) \ldots (i + k))$. For example, (123) is a three-cycle mapping a_1 to a_2, a_2 to a_3, and a_3 to a_1.

Remark 2.1 We see that a permutation of a set S is simply a rearrangement of the members of the set S, and any rearrangement of the members of S is a permutation of S.

Definition 2.2 If S_1, S_2, and S_3 are sets, $f_1 : S_1 \to S_2$ is a function, and $f_2 : S_2 \to S_3$ is a function, then their *composition* $f_2 \circ f_1$ is defined to be the function from S_1 to S_3 given by $f_2 \circ f_1(x) = f_2(f_1(x))$, for $x \in S_1$.

Proposition 2.1 *Let S be any set and f_1, f_2, f_3 any permutations of the set S.*

(i) *$f_1 \circ f_2$ is a permutation of S;*
(ii) *$f_1 \circ (f_2 \circ f_3) = (f_1 \circ f_2) \circ f_3$ (associativity);*
(iii) *if $f_0(x) = x$, for all $x \in S$, then f_0 is a permutation of S and is called the identity permutation (existence of identity);*
(iv) *there exists a permutation $f^{\leftarrow}$ of S such that $f_1 \circ f_1^{\leftarrow} = f_1^{\leftarrow} \circ f_1 = f_0$. The permutation $f_1^{\leftarrow}$ is called the inverse permutation (existence of inverse).*

Proof. Exercise. □

Remark 2.2 can be ignored if you have not previously studied abstract algebra or group theory. It is not needed in our study of probability.

2.2 Card Games

Remark 2.2 We saw in Proposition 2.1 that the set G of all permutations of a given set S has a binary operation $\circ$ acting on any two permutations such that G with this binary operation has the propertiy that $f_1 \circ f_2 \in G$, whenever $f_1, f_2 \in G$, that $\circ$ is an associative operation, that G has an identity, and each member of G has an inverse. If you have studied abstract algebra or group theory, you will recognize that this tells us that G is a *group* with the group operation being $\circ$. G is called a *permutation group*. We note further that $\circ$ is not a commutative operation; for example, consider the set $\{1, 2, 3, 4\}$, f_1 is the cycle (123) and f_2 is the cycle $(2, 4)$, then $f_1 \circ f_2 \neq f_2 \circ f_1$, since $(24)(123) = (2,431) \neq (123)(24)$. So G is not a so-called abelian group, that is, G is not a commutative group.

Proposition 2.2 *If S is a finite set with $n \in \mathbb{N}$ members, then there are $n!$ distinct permutations of the set S.*

Proof. Let $S = \{a_1, a_2, \ldots, a_n\}$ and $f : S \to S$ a permutation. Observe that $f(a_1)$ equals one of $a_1, a_2, \ldots, a_n$. So there are n different possibilities for $f(a_1)$. As f is one-to-one and onto, there then are $n - 1$ different possibilities for $f(a_2)$. And $n - 2$ possibilities for $f(a_3)$ and so on. So altogether there are $n \times (n - 1) \times \cdots \times 1 = n!$ different possibilities for f. □

Remark 2.3 It is very important that we understand set theory notation. When we say that $S = \{a_1, a_2, \ldots, a_n\}$ is a set, we mean

(i) the set S has n distinct members, that is, $a_i \neq a_j$, for $i, j \in \{1, 2, \ldots, n\}$ and $i \neq j$; and
(ii) the order $a_1, a_2, \ldots, a_n$ is immaterial. For example,

$$\{a_1, a_2, \ldots, a_n\} = \{a_n, a_{n-1}, \ldots, a_1\} = \{a_2, a_3, a_1, a_4, a_5, a_6, \ldots, a_n\}.$$

So while Proposition 2.2 tells us that the set S has $n!$ permutations, all of these $n!$ sets are equal as sets.

2.2 Card Games

There seems no more appropriate way to begin the section on card games than with a quote from Oscar Fingal O'Fflahertie Wills Wilde (1854–1900).

> One should always play fairly when one has the winning cards.
> — Oscar Wilde

The 52-Card Deck

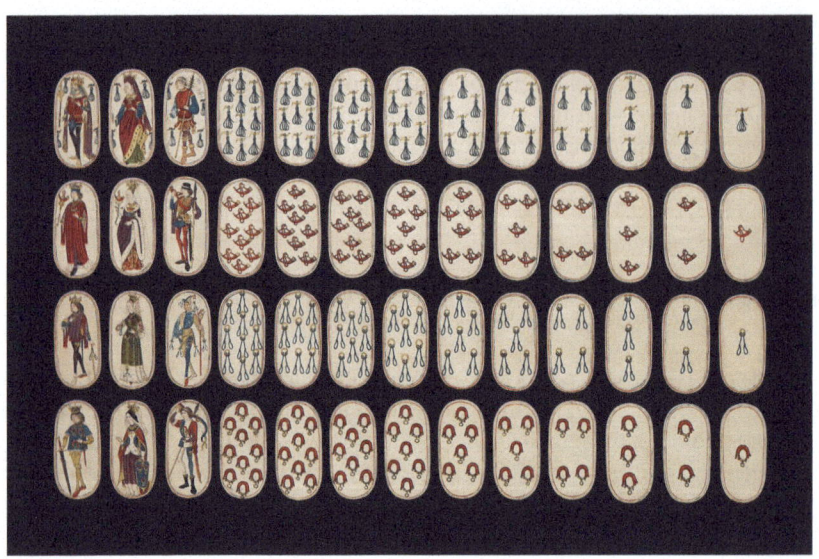

The Cloisters set of 52 cards that constitutes the only known complete deck of illuminated ordinary playing cards (as opposed to tarot cards) from the fifteenth century is owned by the Metropolitan Museum of Art in New York. It is thought that it was manufactured in the 1470s. The cards are hand drawn with highlights of gold and silver in the technique of the time used for illuminated manuscripts. There are four suits, each consisting of a king, queen, knave, and ten pip cards. The suit symbols, based on equipment associated with the hunt, are hunting horns, dog collars, hound tethers, and game nooses. There are no jokers.

The Standard 52-Card Deck

We saw in Example 1.11 the standard 52-card deck of cards, which is the French-English design, most commonly used, especially in the English-speaking world. Manufacturing these cards in England began in the sixteenth century. By the nineteenth century, the English pattern had spread throughout the world. Some decks have bar code markings to facilitate sorting by machines. The standard deck uses black for spades and clubs and red for hearts and diamonds. However some decks use four colours.

2.2 Card Games

The Origin of the Sandwich

It is suggested that John Montagu (1718–1792), fourth Earl of Sandwich, an eighteenth-century English aristocrat while playing a game of the card game cribbage did not want to interrupt his game to eat a meal. So he ordered his valet to bring him salt beef between two pieces of toasted bread. This allowed him to continue playing while eating without getting his hands and the cards greasy from eating meat with his bare hands. The dish grew in popularity in London, and the name of Lord Sandwich became associated with it. Of course throughout history, all sorts of food had been put inside wraps, especially in the Middle East and India. In the Jewish Passover tradition, the so-called Hillel sandwich dates back 2,000 years.

Cribbage, Noddy, and Costly Colours

We begin our discussion with one of the oldest card games played with the 52 card deck, namely, the game of *noddy*. It is referred to in the Oxford English Dictionary of 1589. It can be thought of as a precursor of the game of *cribbage*. While noddy means fool or simpleton, here it refers to knave—a person of humble birth or position. It was played with the ace being the low card (rather than the high card or as a choice of either high or low as occurred in later games). As this game appears to be extinct, we shall not discuss it further. A probable descendant of noddy is the English game of *Costly Colours* which by 1850 was described as obsolete. The interested reader may consult [22, 29].

Sir John Suckling: Inventor of Cribbage

Sir John Suckling (1609–1641) was an English poet who is credited with having invented the card game *cribbage*. He based cribbage on the game noddy. He is said to have sent numerous packs of marked playing cards to aristocratic houses in England and then travelled around playing Cribbage with them. Apparently he won about 20,000 pounds, the equivalent of US$6 million today. He had anything but a dull life: he assisted King Charles 1 in the first Scottish war, raising a troop of a hundred men at very considerable personal cost and acquitted himself well as a soldier. He was elected to Parliament but fled to France when found guilty of high treason because of an attempt to restore the power of the King over Parliament. He later eloped to Spain where he fell into the hands of the Inquisition. Finally he committed suicide by poison in Paris for fear of poverty. See [1].

Muggins in the English Language

Perhaps you have referred to yourself as a *muggins* when you did something foolish or volunteered for some role or job unnecessarily. In English the term means "a foolish and gullible person (often used humorously to refer to oneself)". Few would know that it comes from the game of *Cribbage*. If a Cribbage player fails to claim their full score on any turn, their opponent may call out "Muggins" and peg any points overlooked by the player.

Cribbage in Literature, Movies, TV, and Video Games

The game of *cribbage* was immortalized in the 1840 novel *The Old Curiosity Shop* written by Charles John Huffam Dickens (1812–1870). It is also mentioned in the following two centuries in the 1973 movie *The Sting* with Paul Newman and Robert Redford playing cribbage; in a scene in the popular TV series M*A*S*H (1972–1983) in which the smug and arrogant doctor Major Charles Emerson Winchester III is playing Cribbage; and in the 2010 video game *Red Dead Redemption*.

Self-Proclaimed Cribbage Capital of the World: Nelson, Montana

The earliest rules for cribbage were published in England in 1662. Cribbage was popular amongst prospectors in the American West and the small mining town of Nelson in the state of Montana still has a sign asserting that it is the cribbage capital of the world. Cribbage was also popular with submariners.

Cribbage Game Rules as Explained by ChatGPT

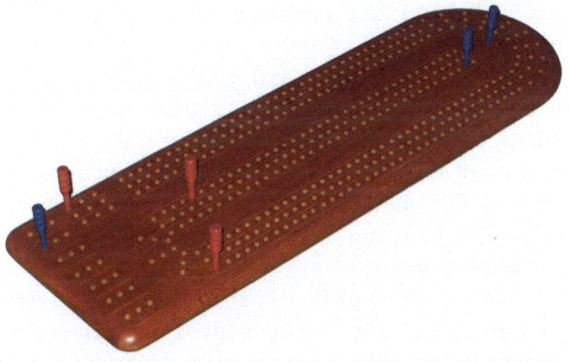

As an experiment I asked ChatGPT to summarize the rules of *Cribbage*. I let readers judge how well this is done.

```
Cribbage is a card game that involves two players,
a standard 52-card deck, and a distinctive scoring board.
Here are the basic rules:

### Objective:
The goal is to be the first player to score 121 points,
usually tracked on a cribbage board.

### Setup:
1. Each player cuts the deck to determine who deals first.
The player with the lower card deals.
2. The dealer shuffles and deals six cards to each player.
3. Each player selects two cards to place face down into
the crib, forming a four-card hand for the dealer to
score later.

### Play:
```

1. The non-dealer cuts the remaining deck, and the dealer reveals the top card of the bottom portion, called the "starter" or "cut" card. If this card is a Jack, the dealer scores two points ("his heels").
2. Players take turns playing one card at a time, announcing the running total of card values. Face cards are worth 10 points, Aces are worth 1 point, and all other cards are worth their face value.
3. The goal is to create specific combinations:
    - **15:** Making the running total exactly 15 scores 2 points.
    - **31:** Making the running total exactly 31 scores 2 points.
    - **Pairs:** Playing a card that makes a pair (same rank as the previous card) scores 2 points.
    - **Runs:** Playing cards that form a run (sequence of three or more consecutive cards) scores points equal to the number of cards in the run.
    - **Flush:** Playing four cards of the same suit in a row scores 4 points.
4. If a player cannot play a card without exceeding 31, they say "go," and the other player continues until they also cannot play. The last player to play a card scores 1 point ("go"), or 2 points if they make 31 exactly.
5. After all cards are played, players score their hands.

Scoring Hands:
1. Players score their hands and the crib, counting combinations of cards:
    - **15:** Each combination of cards that adds up to 15 scores 2 points.
    - **Pairs:** Each pair of cards of the same rank scores 2 points.
    - **Runs:** Sequences of three or more consecutive cards score points

Scoring Hands (continued):
   - **Flush:** If all four cards in the player's hand are of the same suit, it scores 4 points. If the starter card is also of the same suit, it scores 5 points. In the crib, the flush must include the starter card to count and scores 5 points.
   - **Nobs:** If a player has a Jack of the same suit as the starter card, it scores 1 point ("one for his nobs").

2.2 Card Games

Example of Scoring a Hand:
Let's say a player's hand consists of the 5S, 5C, 6D, and 7H, with a
starter card of 8D:
- 15s: 5S + 5C + 5 combinations = 6 points (three different 15s)
- Pairs: 5S + 5C = 2 points
- Runs: 5S, 6D, 7H = 3 points
- Total: 11 points

Scoring the Crib:
After both players have scored their hands, the dealer scores the crib
using the same combinations as above. The dealer scores both the
non-dealer's hand and the crib, but the non-dealer only scores their hand.

Winning the Game:
The game continues with players alternating the deal until one player
reaches 121 points. If a player reaches 121 points during the play or
scoring of hands, the game ends immediately.

Additional Rules:
1. **Muggins (optional):** If playing with muggins, players can claim
   points that their opponent missed during scoring.
2. **Skunks:** Winning by 30 or more points
   (e.g., scoring 121 points to your opponent's 90 or less) is called a "skunk"
   and is sometimes counted as a double win.

Summary:
- **Dealing and discarding:** Players receive six cards and discard two to the crib.
- **Play phase:** Players alternate playing cards to reach specific totals
  (15, 31) or combinations (pairs, runs).
- **Scoring phase:** Each player scores their hand, followed by the
  dealer scoring the crib.
- **Winning:** The first player to reach 121 points

> wins the game.
>
> Feel free to ask if you have any more specific
> questions or need further clarification on any
> of these rules!

The reader who is not satisfied with this ChatGPT summary of the rules for cribbage is welcome to consult [20–22, 25].

I mention that cribbage can be played by two to six players, whereas the description above is clearly for six players.

If one wants to know what is the *perfect hand in cribbage*

> In Cribbage, the perfect hand is worth 29 points.
> This rare hand consists of:
> [5Hearts,5Spades,5Clubs,5Diamonds,JackHearts]
>
> - Three fives (5S, 5C, 5D)
> - One five of the same suit as the starter card (5H)
> - The jack of the same suit as the starter card (JH)
>
> The points are calculated as follows:
> - Fifteen counts: Four different combinations of two
>   cards that make 15, and each five with the starter
>   5H = 8 points
> - Pairs: Six pairs of fives = 12 points
> - Four-of-a-kind: All four fives = 4 points
> - Nobs: The jack of the same suit as the starter
>   card = 1 point
>
> Total = 29 points.

I leave as an exercise for you to show that the probability of a perfect hand in a 2-person Cribbage game is $\frac{1}{216{,}580} = 0.000004617\ldots$.

In a three-person cribbage game, each person is dealt only five cards and discards only one in the crib. I leave as an exercise for you to show that the probability of a perfect 29 point hand in a 3-person cribbage game is $\frac{1}{649{,}740} = 00000153908\ldots$.

2.2 Card Games

Poker

According to [29] in 1973, *poker* was by far the most popular card game throughout the world measured both by the amount of money that changes hands daily and by the number of players. He goes on to say that while 30 years earlier Poker was almost exclusively a man's game, by the 1970s more women than men played mainly because so many new poker variations had sprung up over the previous 20 years. Scarne says that the earliest reference to Poker he found was in the writings of Jonathan Harrington Green (1813–1887) [8] who is best known as a reformed American gambler who campaigned against illegal gambling and was responsible for some antigambling laws.

> Today Poker is played with a standard deck of 52 cards. Each player is initially dealt five cards. We shall see that in a game of poker, I am dealt five cards one at a time. As a player I am interested only in what five cards I receive, not in the order that I receive the five cards. So while there are 5! = 120 different orders that I can be dealt those cards, as far as I am concerned, they are all the same.

We shall discuss various poker hands in due course, but we need to cover some basic material first.

Combination Lock

Example 2.2 Let us consider the following five-digit combination lock.

To open this lock, you set each of the five dials to the correct number between 0 and 9, so obtaining a number between 0 = 00000 and 99,999. Thus there are 100,000 choices. Clearly the correct order of these five numbers is essential. Here the order of the numbers does matter.

Selection with and Without Replacement

Definition 2.3 Let S be a set. An element a is said to be *selected* from S if $a \in S$. Let $k \in \mathbb{N}$. Then the elements a_i, $i = 1, 2, \ldots, k$, are said to be *selected without replacement* or a *sample without replacement* from S if a_1 is selected from S and each a_i, for $i = 2, \ldots, k$, is selected from the set $S \setminus \{a_1, a_2, \ldots, a_{i-1}\}$.
The elements a_i, $i = 1, 2, \ldots, k$, are said to be *selected with replacement* or a *sample*

with replacement from S if each a_i, for $i = 1, 2, \ldots, k$, is selected from the set S. (In a sample with replacement, the same element may appear more than once.)

Example 2.3 Let us begin with a pack of 52 cards as in Example 1.11 and assume that we are dealt five cards as in a common version of the game *poker*. Then this is an example of *selection without replacement*, so the five cards must be distinct. If Alex is playing this game, then the card Alex receives first could be any one of the 52 cards, her second card can be any one of the remaining 51 cards, her third card can be any of the remaining 50 cards, her fourth card can be any of the remaining 49 cards, and her fifth card can be any of the remaining 48 cards. So the number of these *ordered* five card hands is $52 \times 51 \times 50 \times 49 \times 48 = \frac{52!}{47!} = 311{,}875{,}200$. But this does *not* tell us the number of possible hands Alex can be dealt. This is because the order that Alex receives her five cards is of no consequence: for example, if she is dealt AH,2H, 3H, 4H, 5H, or 2H,AH,3H,4H,5H, then she has the same five cards. By Proposition 2.2 there are 5!=120 different orderings (permutations) of these five cards. So the number of distinct five card hands is $\frac{52!}{(47!)(5!)} = 2{,}598{,}960$.

```
Using the R language
factorial(52)/(factorial(47)*factorial(5))
```

Proposition 2.3 *Let the set S have n elements, for $n \in \mathbb{N}$, and let $k \in \mathbb{N}$ with $k \le n$. Consider samples from S consisting of precisely k elements.*

 (i) *There are n^k distinct samples, if each is a sample with replacement from S.*
 (ii) *There are $n(n-1)(n-2)\ldots(n-k+1) = \frac{n!}{(n-k)!}$ distinct samples, if each is a sample without replacement from S.*

Proof. When sampling without replacement, each of the k elements can be chosen in n ways. Thus (i) is true.

When sampling without replacement, the first element can be chosen in n ways, the second element can be chosen in $n-1$ ways, and so on yielding that (ii) is true. □

Definition 2.4 Let $n, k \in \mathbb{N}$, with $k \le n$, and let S be a set which has n members. Then P_k^n denotes the number of ways that an ordered collection $a_1, a_2, \ldots, a_k$ can be selected without replacement from S.
A *combination* is a selection without replacement of k elements from the set S, where the order of selection is unimportant. Further, the number $\binom{n}{k}$, called *n choose k*, denotes the number of combinations obtained by choosing k elements from the set S or equivalently the number of ways of choosing k distinct elements from S or equivalently the number of subsets $\{a_1, a_2, \ldots, a_k\}$ of S.
By convention $\binom{n}{0}$ is defined to be equal to 1.

2.2 Card Games

Using the method in Example 2.3, we can easily prove the following general result.

Proposition 2.4 *Let $k, n \in \mathbb{N}$ with $k \leq n$. Then*

(i) $P^n_k = \frac{n!}{(n-k)!}$;

(ii) $\binom{n}{k} = \frac{n!}{(n-k)!k!}$; *and*

(iii) $P^n_k = k!\binom{n}{k}$.

Proof. Exercise. □

So the number of ways of selecting k items from n items without replacement

(i) when the order is important is $P^n_k = \frac{n!}{(n-k)!}$, and

(ii) when the order is irrelevant is $\binom{n}{k} = \frac{n!}{(n-k)!k!}$.

Example 2.4 We saw in Example 2.3 there are $\frac{52!}{(47!)(5!)} = 2{,}598{,}960$ distinct five card poker hands. We shall assume that the game is fair, so that the probability of being dealt any one card is equal to the probability of any other card.

What, then, is the probability of being dealt four aces amongst the five cards? So 4 of the 5 cards must be AH,AD,AC,AS, and the other card can be any of the remaining 48 cards. So the number of combinations of 5 cards which have 4 aces is 48. As the game is fair, the probability of this occurring is

$$\frac{48 \times (47!)(5!)}{52!} = \frac{(48!)(5!)}{52!} = \frac{1}{13 \times 17 \times 5 \times 49} = 0.00001846\ldots.$$

Example 2.5 In Example 2.4 we calculated the probability of being dealt four aces amongst our five cards. What is the probability if instead of four aces we insisted on "four of a kind", that is, four aces or four kings or four jacks or four 2s or four 3s ... four 10s. We saw the number of combinations of 5 cards which include 4 aces is 48. This is also true if four aces is replaced by four kings or four queens ... four 2s. So the number of combinations with four of a kind is 13×48. So the probability of being dealt four of a kind is

$$13 \times \left(\frac{1}{13 \times 17 \times 5 \times 49}\right) = \frac{1}{17 \times 5 \times 49} = 0.0002400\ldots.$$

Example 2.6 In some sense we calculated the probability of being dealt four aces and being dealt four of a kind from first principles in the previous examples. Let us do it now using $\binom{n}{k}$. First, let us calculate the probability of being dealt four aces in a fair game. Note that the deck of 52 cards has 4 suits: spades(S), clubs(C), hearts(H), diamonds(D), and 13 of each suit, A,K,Q,J,10,9,...,2. Now four cards of those dealt are aces, and these four cards can be chosen in only one way. For the fifth card, we choose any of the four suits, so that is $\binom{4}{1}$. For the value of the card, it is any of the 13 cards except the ace, so it is $\binom{12}{1}$. So the total number of combinations for being dealt four aces is $\binom{4}{1} \times \binom{12}{1} = 4 \times 12 = 48$. As before the probability is then this number divided by the total number of distinct five card hands, that is,

$$\frac{48 \times (47!)(5!)}{52!} = \frac{5!}{52 \times 51 \times 50 \times 49} = \frac{1}{13 \times 17 \times 5 \times 49} = 0.00001846\ldots$$

Next let us look at the number of combinations of five card hands which have four of a kind. This time the 4 of a kind can be any of the 13 values, that is, $\binom{13}{1} = 13$. The 5th card can be any of the remaining 12 values and any of the 4 suits, that is $\binom{12}{1} \times \binom{4}{1} = 48$. So the total number of combinations with 4 of a kind is 13×48. Thus the probability of being dealt four of a kind in a fair game is

$$13 \times \left(\frac{1}{13 \times 17 \times 5 \times 49}\right) = \frac{1}{17 \times 5 \times 49} = 0.0002400\ldots$$

Example 2.7 Again let us consider a fair game of five-card poker. We evaluate the probability that the five cards we are dealt, ignoring the suits, have five different face values. The face values can be chosen in $\binom{13}{5}$ ways. Each of the cards can be any one the four suits. So the total number of combinations is $4^5 \times \binom{13}{5}$. The probability of this happening is the number of combinations divided by the total number of five card hands which is $\binom{52}{5}$, that is, the probability is

$$\frac{4^5 \times \binom{13}{5}}{2{,}598{,}960} = 0.5070\ldots$$

We evaluated the above using the software package R:

```
(4^5*choose(13,5))/choose(52,5)
```

2.2 Card Games

Example 2.8 Consider a fair game of five-card poker. We evaluate the probability of the five cards we are dealt being a *full house*, that is, three cards of one value and two cards of another value but equal value to each other.

First examine the three cards of a kind. There are 13 different values, so the number of ways of choosing the value is $\binom{13}{1}$. Now there are four suits, and the three cards are therefore of three suits, giving the number of combinations of these as $\binom{4}{3}$. So the total number of combinations of three cards of equal value is $\binom{13}{1} \times \binom{4}{3} = 52$.

We also need a pair of a different value. This pair can be any of 12 values, so there are $\binom{12}{1}$ combinations. The suits are any two of the four possible suits, that is, we have $\binom{4}{2}$ combinations. So in all we have $\binom{12}{1} \times \binom{4}{2} = 72$ combinations of the pair. So altogether the number of combinations of the five cards for a full house is $52 \times 72 = 3,744$.

We know previously that the total number of 5 card hands is 2,598,960. So the probability of a full house in our fair game is

$$\frac{3,744}{2,598,960} = 0.001440\ldots.$$

Example 2.9 Consider a fair game of five-card poker. We evaluate the probability of the five cards we are dealt being a *flush*, that is, all five cards are from the same suit. (In the special case that the five cards have the values ace, king, queen, jack, and 10, the flush is called a *royal flush*.)

All the cards must be of the same suit, and there are four possible suits. There are 13 different values, and we are dealt 5 cards, so the total number of 5-card flushes is $\binom{13}{5} \times \binom{4}{1} = 5,148$. The total number of 5 card hands is 2,598,960. So we see that the probability of being dealt a flush is

$$\frac{5,148}{2,598,960} = 0.001980\ldots.$$

Example 2.10 Consider a fair game of five-card poker. We evaluate the probability of the five cards we are dealt being a *straight*, that is, the values of the five cards are consecutive numbers, and they can be of any suits. (In the special case that the five cards are of the same suit, the straight is called a *straight flush*.)

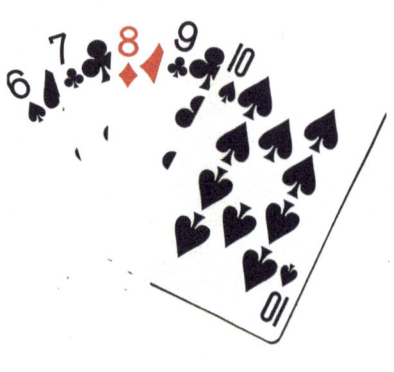

We note that in a straight the lowest value card can be ace, 2, 3, 4, 5, 6, 7, 8, 9, or 10. Note also that {ace, 2, 3, 4, 5} and {10, jack, queen, king, ace} are both straights.

There is no straight which has as its lowest value card a jack, queen, or king. Now once we know the lowest value card in a straight, we know the values of the higher value cards in the straight. What we do not know is the suits of each of these cards. For each card in the straight, there are four possible suits. So for each of ace,2,3,4,5,6,7,8,9,10, there are 4^5 straights which have that value card as its lowest.

Thus there are $10 \times 4^5 = 10{,}240$ different straights. Altogether there are 2,598,960 possible hands. So the probability of your hand being a straight in a fair game is

$$\frac{10{,}240}{2{,}598{,}960} = 0.003940\ldots.$$

We see, therefore, from Example 2.9, that there is a higher probability of a straight than a flush. Indeed the probability of a straight is almost twice that of the probability of a flush.

Bluffing in Poker

To quote Judi James in her book *Poker Face: Mastering Body Language to Bluff, Read Tells, and Win*, [12], "Winning at poker is like winning at life. To play well you need more than a lucky hand. That exciting extra dimension is all about people skills; having the ability to read your opponent's body language and block their attempts to read yours". Of course this is the source of the term *poker face* in English idiom. I recommend reading [12] to learn how to bluff well in poker.

I also recommend consulting the book [32] which deals not only with bluffing but playing poker competitively, in general.

Finally I mention the substantial book *Decide to Play Great Poker: A Strategy Guide to No-Limit Texas Hold'Em*, [7], by Annie Duke and John Vorhaus.

2.2 Card Games

Bridge

Example 2.11 The card game *whist*, [20], introduced in the eighteenth century is a descendent of the sixteenth-century game *trump*. In 1742 Edmond Hoyle (1672–1769) published the book, [10], which became a standard text on the game until 1862 when Henry Jones, writing under the pseudonym of Cavendish, published [14]. In fact the phrase "according to Hoyle" came to mean playing by the rules.

During the 1890s a variant of the game known as *bridge whist* appeared and it evolved into *contract bridge*. For a discussion of contract bridge and probability, see [19]. Whist is an example of a *trick-taking game*, that is, one where players take turns playing cards, and the highest card played in each round wins what is known as the *trick*.The goal is to win more tricks than others. this family of games includes contract bridge, *euchre*, and *five hundred*.

Below is a drawing by Marguerite Martyn (1878–1948) for the St. Louis Post-Dispatch of a session of the Women's Whist Club Congress in St. Louis, Missouri.

Charles Anderson Worsley Anderson-Pelham, second Earl of Yarborough (1809–1862), gave his name to a specific hand of cards in *whist* and *contract bridge* which is known as a *Yarborough*. In the game of whist, he would offer to give any player £1,000 (the equivalent in 2020 of over £100,000 or over $100,000) if during the evening they were dealt a hand that contained no card higher than a 9. All he asked was that the player pledged him £1 before each deal. Let us look at the probability of him losing.

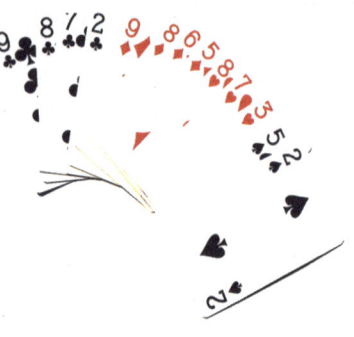

Bridge, today, is one of the most popular card games in the world. Whist and contract bridge are played by four players in two competing partnerships with partners usually sitting opposite each other around a table. One pack of 52 cards is used and each player is dealt 13 cards. In bridge, one pair is referred to as North and South, while the other is West and East. The game begins with an auction, the details of which are not relevant to our study. Players may not see their partner's hand during the auction. The dealer begins the auction and can make the first bid. Each player in turn may then pass or make a bid—a contract. The bid specifies the level of their contract, which must be higher than any previous bid including their partner's and may specify the trump suit or no trump (no denomination). The term "higher" means the value of the card is higher or the suit is regarded as higher, where the suits have the following order from low to high: clubs, diamonds, hearts, spades, and no trump.

So firstly, let us look at the probability of one player's hand of 13 cards being a Yarborough. The number of hands which are a Yarborough is the number of hands using only cards from 2, 3, 4, 5, 6, 7, 8, 9. There are 32 such cards and 13 cards are selected from these 32. So the number of Yarborough hands is $\binom{32}{13}$. The total number of bridge or whist hands is $\binom{52}{13} = 635{,}013{,}559{,}600$ (over 635 billion hands).

So the probability of a Yarborough is $\binom{32}{13}/\binom{52}{13}$. To calculate the odds we evaluate

$$\binom{52}{13}/\binom{32}{13} = 1{,}828.042.$$

> We used the software package R:
> choose(52,13)/choose(32,13)

So the correct odds are >1,827:1. We now know that the 2nd Earl of Yarborough was on a good deal since he offered odds of only 1000:1.

Let us now look at the probability of more common bridge hands. There are four suits, and so a bridge hand could have, for example, 4-4-3-2 suits, that is, four cards in each of two suits, three cards in a third suit, and two cards (a doubleton) in the fourth suit. Let us evaluate the probability of this 4-4-3-2 hand by calculating the number of such hands. Note that 4-4-3-2 does not specify which suits have 4, 3, or 2 cards. If we specify which suit has the doubleton and which suit has three cards, then we know the other two suits have four cards. The suit having the doubleton can be chosen in four ways, and the suit having three cards can then be chosen in three ways. So the total number of hands of the type 4-4-3-2 is

$$\binom{13}{4} \times \binom{13}{4} \times \binom{13}{3} \times \binom{13}{2} \times 4 \times 3.$$

The probability of a 4-4-3-2 hand is therefore this number divided by $\binom{52}{13}$. This is calculated using the software package R and equals $0.2155\ldots$.

This is, in fact, the highest probability of all possible hands.

Similarly we can calculate the probability of each of the other 39 hands:

5-3-3-2, 5-4-3-1, 5-4-2,2, ..., 12-1-0-0, 13-0-0-0.

For example, the probability of 13-0-0-0 is $4/\binom{52}{13} = 0.000000000006299\ldots$.

We might ask what is the probability that one player and her/his partner both have 4-4-3-2 hands. This is left as an exercise.

Bidding in Bridge

To quote Andrew Robson and Oliver Segal in their book *Partnership Bidding at Bridge: The Contested Auction*, [27], "What makes bridge interesting to you? Why does it engage you sufficiently that you are reading a book about the game? Of course bridge offers an endless mental challenge—but then so does chess, backgammon, or crossword puzzles for that matter, no; you probably play bridge for much the same reason that we do– because it is essentially and uniquely a partnership game. ... Bridge is a game at which two people should *combine* rather than add their efforts. ... it is clearly in the auction (the bidding stage) where most of the opportunities lie to exploit partnership skills". Robson and Segal say that their book [27] is written so that life will be easier for your partners.

One of the best books for beginners to bridge is Crisfield's book *Bridge for Everyone*, [5]. It is very gentle and holds your hand as you learn bridge basics, scoring, bidding, and playing.

The book *Basic Bridge: The Guide to Good ACOL Bidding & Play*, [16], by Ron Klinger can also be read by the beginner. It focusses however on bidding. It says "A bidding system is like a language—it is a means of communicating with your partner. However the language of bridge allows only 15 legal words: one, two, three, four, five, six, seven, no-trumps, spades, hearts, diamonds, clubs, double, redouble, and pass (no-bid). ... With this restricted language you try to convey to your partner your thirteen cards ... " of a possible billion possibilities. "Basic Bridge is based on the ACOL system, one of the most popular systems in the world. it is the easiest to learn and is the most natural of all bidding systems".

The book *Insights on Bridge: Moments in Bidding, Book 1*, [17], by Mike Lawrence provides over 100 examples of bridge hands and how bidding might proceed.

Euchre

Euchre (or *eucre*) was regarded as the national card game of the United States in the late nineteenth century. According to [34] "During the reign of Napoleon in Europe, Euchre was modernized and brought to America in the French controlled New Orleans. From Louisiana the game travelled up the path of the Mississippi River into the northern states where it gained considerable popularity". It is a *trick-taking* card game played to this day in Australia, Canada, New Zealand, the UK, and the USA. There are many variants of the game of euchre. For details on how to play euchre, see [2–4, 29].

Euchered lithograph 1884

Euchre was responsible for introducing the *joker* into the modern deck of cards. Typically *euchre* is played by 4 players, and each is dealt 5 cards from a 25-card deck consisting of ace, king, queen, jack, 10, 9 in each suit plus one Joker. Having decided on a trump suit—let us say it is spades, the cards rank downwards:

1. Joker
2. Jack of spades
3. Jack of clubs (the jack in the other black suit)
4. Ace-king-queen-10-9 of spades (in that order)
5. Ace-king-queen-jack-10-9 of other suits (except jack of clubs which already appears above).

Joker

Example 2.12 In the game of euchre, as described above, what is the probability of being dealt two Jacks of the same colour and three other cards (not the Joker) of the same suit as the first Jack you were dealt. (While the euchre hands do not depend on the order the cards were dealt, we can nevertheless observe that order and so referring to the first jack dealt makes sense.)

Firstly we observe that the total number of different euchre hands is $\binom{25}{5} = 53,130$. As there are four jacks, there are four ways of choosing the first Jack and then only one way of choosing the second jack of the same colour. So there are $4 \times 1 = 4$ ways of choosing the two jacks of the same colour. The other three cards are then chosen from ace, king, queen, 10, 9, and they must all be of the suit as the first jack dealt. So of these five cards, we are dealt three. There are $\binom{5}{3} = 10$ ways this can occur. So the number of ways we can be dealt a hand as described is $4 \times 10 = 40$. So the probability of being dealt such a hand is

$$40/53,130 = 0.0007528\ldots.$$

2.2 Card Games 75

Example 2.13 If in the game of euchre, as above, spades is the trump suit, what is the probability of being dealt the highest possible scoring hand?
The best hand is obviously joker, jack of spades, jack of clubs, ace of spades, and king of spades. There is only one such hand. So the probability of this is

$$1/53{,}130 = 0.00001882\ldots.$$

Rummy

Playing Conquian in New Orleans, 1938

Example 2.14 Rummy evolved from a Spanish game called *conquian* and appeared in southwest USA in the 1850s. There are a hundred or more games in the rummy family. In the 1930s and 1940s, *rummy*, [21], was the most popular card game in the USA until its offsprings *gin rummy* and *canasta* [20] became more popular. There are many variants of the game of Rummy which can be played by two to six people. We shall focus on basic rummy with 4 players, where each player is dealt 7 cards from a standard 52-card deck, with the cards ranking from 2 (low) to ace (high). (Some variants have a 54-card deck which includes 2 jokers, and each joker is a wild card which can be used instead of any other card.) All rummy games are concerned with *Melds*, that is, a combination of sequences (or *Runs* of 3 or more cards in the same suit and/or *Sets* (or *Books*) of 3 or 4 of a kind as in the pictures below.

Players pick up and discard a card when it is their turn. Melds are placed face up on the table. The player who successfully melds all her/his cards before everyone else is the winner, scoring points based on the value of the cards held in the hands of the other players.

Now let us look at some of the probabilities. There are 52 cards, and each player is dealt 7 cards. So the total number of different rummy hands is $\binom{52}{7} = 133{,}784{,}560$.

Now consider being dealt four of a kind. Each suit has 13 cards, so the total number of 4 of a kind is 13. The other 3 cards in the hand can be any of the other 48 cards. So there are $\binom{48}{3} = 17{,}296$ ways of choosing those 3 cards. Thus the probability of being dealt 4 of a kind is

$$\frac{13 \times 17{,}296}{133{,}784{,}560} = 0.001680\ldots.$$

We leave as an exercise calculation of the probability of being dealt (i) three or four of a kind (ii) a sequence of at least four cards (iii) a sequence of at least three cards (iv) a hand which consists of a four card run and a three card set.

There are many books on rummy. A useful one is Trev Tobin's *How Do I Play Rummy: The Ultimate Guide to Learn How to Play, Master and Win Rummy, Instructions, Rules, and Strategies to Excel at Playing Rummy*, [31]. This book, available in Kindle and paperback, also includes how to play gin rummy.

2.3 Mahjong

The game *mahjong* (or *mah jongg*) is a tile-based game belonging to the rummy family. It was probably developed during the Qing dynasty (1644–1911) in China (although some argue that it dates back 2,600 years to Confucius) and has spread throughout the world since the early twentieth century. According to [35], mahjong was so popular in the 1920s that people often paid for teachers and for real-time practice, and it was common for husbands to teach wives and for adults to teach children. There is anecdotal evidence suggesting participation can prevent mental deterioration associated with old age.

Mah Jong

2.3 Mahjong

There are variants of mahjong such as *Hong Kong Mahjong, Harbin Mahjong, Tianjin Mahjong* (which has seven jokers), *Japanese Mahjong, American Mahjong (Mah Jongg), Rummikub (Israel Mahjong), Vietnamese Mahjong*, and *Pussers Bones* played by sailors in the Royal Australian Navy (pusser is slang for a purser or a ship's supply officer), which are played with differing tiles and a differing number of tiles. There is also a related one person game of *Mahjong Solitaire (Shanghai Solitaire)* and a *Three Player Mahjong*. The earliest surviving mahjong sets date to the 1870s. For a discussion of different versions of the game of mahjong, see [26]. There is no definitive statement on the number of mahjong players throughout the world, but it is probably in the hundreds of millions and likely at least as many women as men.

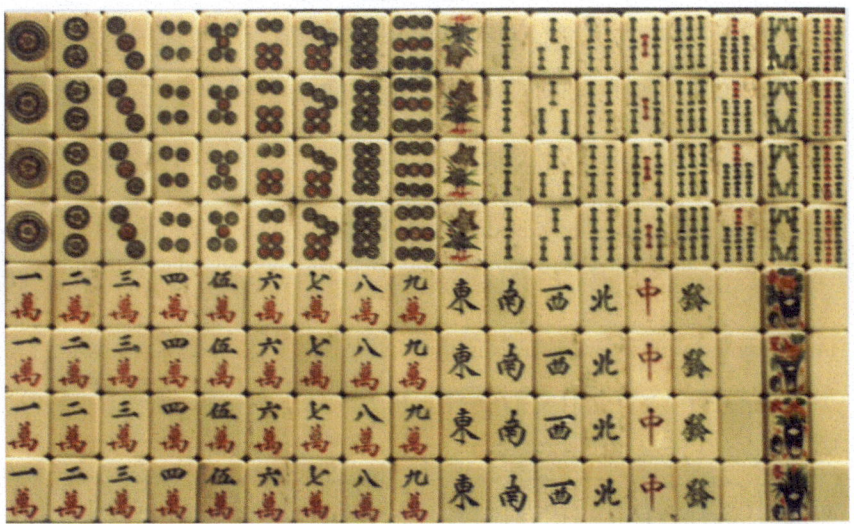

Old Mahjong Tiles Set displayed in the Tianyi Pavilion Museum

We describe a Mahjong game played with 144 tiles with Chinese characters and symbols. Each player begins by receiving 13 tiles. In turn players draw and discard tiles until they complete a winning hand using a 14th tile drawn to form four melds (or sets) and a pair (eye).

The 144 tiles are:

1. 4 identical sets of 9 tiles of dot-style tiles = 36 dot-style tiles;
2. 4 identical sets of 9 tiles of bamboo-style tiles = 36 bamboo-style tiles;
3. 4 identical sets of 9 tiles of character-style tiles = 36 character-style tiles;
4. 4 identical sets of East, South, West, and East wind honor tiles = 16 wind honor tiles;

5. 4 identical sets of centre, fortune, and blank arrow tiles = 12 arrow tiles;
6. Spring, Autumn, Winter, and Summer tiles = 4-season tiles;
7. Plum, orchid, chrysanthemum, and bamboo tiles = 4 flower tiles.

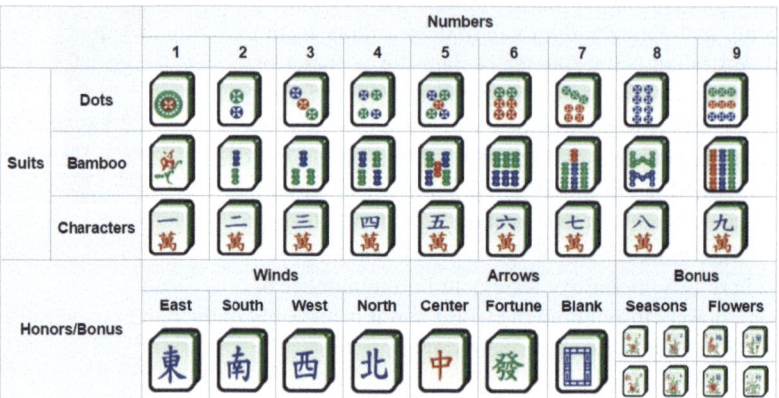
Mahjong Tiles

Melds in mahjong correspond to melds in rummy. *Melds* are sets of tiles in a player's hand consisting of:

(i) a *Pong*, which is three identical tiles, or
(ii) a *kong*, which is four identical tiles, or
(iii) a *Chow*, which is three tiles of the same suit in numerical order, or
(iv) *Eyes* (or an it pair), which are two identical tiles and are an essential part of a 14-tile winning hand.

As examples of these, we have as follows.

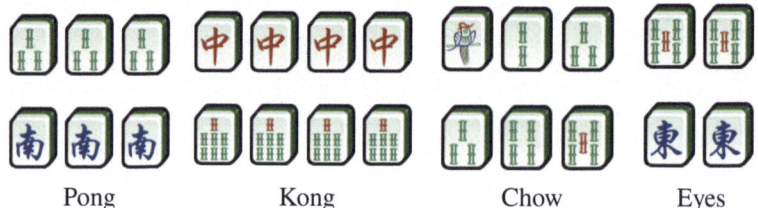
Pong   Kong   Chow   Eyes

Each of the four players is seated in a designated area; tiles are turned face down and moved around by all players to shuffle them. Then each player forms a row of 18 tiles, 2 tiles high and face down in front of himself or herself. There are various rituals which we shall not elaborate on here. (See [18].) Each player has a hand of 13 tiles. Throughout the game tiles are taken and discarded.

If a player draws a flower or season tile, it is placed to the side and not part of the 13-tile hand (but if that player wins, she/he may earn bonus points for these), and a replacement tile is drawn before their discard.

Whenever a kong is formed, the player must draw an extra tile and then discard a tile. The fourth piece of a kong is not considered as one of the 13-tile hand.

2.3 Mahjong

The number of points the player with the winning hand receives depends very roughly on the probability of that hand occurring.

The total number of 14-tile hands (which must exclude all flowers and seasons and all kongs) is ($\binom{136}{14}$)—the number of hands which include a kong. There are 34 possible kongs (nine from dots, nine from bamboo, nine from characters, four from winds, and three from arrows). The number of 14-tile hands which exclude all flowers and seasons but include one or more kongs is ($\binom{132}{10}$)×34 = $1.06099959705168 * 10^{16}$. So the total number of legal 14-tile hands is

$$10^{18} \times 4.239\ldots, \text{ a gigantic number.}$$

[However, we note that there are identical tiles, so this gigantic number of hands includes many identical hands. Each tile in a legal 14-tile hand is one of 4 identical tiles. So the total number of legal 14-tile hands no two of which are identical is ≥ the total number of legal 14-tile hands divided by 4^{14}, which equals

$$10^{10} \times 1.579\ldots, \text{ which is still a gigantic number.}]$$

A winning hand consists of 14 tiles (13 tiles in the hand plus one tile that is picked up.) Such a hand must have four melds and one pair, for example, four pongs and one eyes. The number of possible pongs is $\binom{4}{3} \times 36 = 144$. The number of hands with four pongs is

$$\left(\binom{4}{3} \times 36\right) \times \left(\binom{4}{3} \times 35\right) \times \left(\binom{4}{3} \times 34\right) \times \left(\binom{4}{3} \times 33\right) = 361{,}912{,}320.$$

The eyes can come from any of 34 sets of 4 tiles other than the tiles identical to those in the 4 pongs, i.e., there are $\binom{4}{2} \times 30 = 180$. So the total number of winning hands with four pongs and one eyes is

$$361{,}912{,}320 \times 180 = 10^{10} \times 6.514\ldots.$$

So the probability of winning with four pongs and eyes is

$$(10^{10} \times 6.514\ldots)/(10^{18} \times 4.239\ldots) = 10^{-8} \times 1.536\ldots.$$

In the exercises you will calculate the probability of winning with some other hands.

Learning Mahjong and Its Role in Modern American Culture

One on the least expensive books for learning Mahjong is Garcia Jack's Book *Mahjong for Beginners*, [11], available in paperback and on Kindle and only 41 pages. A more comprehensive book also available on Kindle is Larry Kistler's *Play Smarter and Win More Mahjong: Logic, Strategy and Tactics*, [15].

The most fascinating recent book on Mahjong is by Annelise Heinz, *Mahjong: A Chinese Game and the Making of Modern American Culture*, [9]. A review of this book says "Bold, ambitious, and stunningly detailed ... Heinz methodically describes the entwined transpacific history of the game, its many different players (Chinese, white women, Jewish American mothers, Airforce wives, Japanese Americans, and Chinese Americans), and its production, marketing, and consumerism in China and the United States, the game's cultural evolutions, its cross-over appeal, and importantly, the significance that the game had on the formations of race, gender, ethnicity, and national belonging ... Mahjong is both the star and the setting of a compelling study of American modernity". Heinz also reveals the ways in which women leveraged a game to gain access to respectable leisure. The result was the forging of friendships that lasted decades and the creation of organizations that raised funds for the war effort and philanthropy. No other game has signified both belonging and standing apart in American culture.

American Mahjong or Mah Jongg

American mahjong, known as *mah jongg*, is a variant of Mahjong introduced by Joseph Park Babcock (1893–1949) who decided to increase the interest in America in the game by making it easier for Americans to understand. So he produced new simpler rules that became standard for the American game in 1935. Now national tournaments are played using these rules. The total number of tiles in Mah Jongg is 152. As well as those pictured at the right there are dot tiles, bamboos, craks, windows, dragons, flowers, and jokers.

Elaine Sandberg's book, *A Beginner's Guide to American Mah Jongg: How to Play The Game & Win*, [28], is available in paperback and Kindle. See also [6].

2.4 Lotto

Example 2.15 Throughout the world, there are many lotteries whose names include the word *"lotto"*. For our discussion, let us assume that to win a share of the first prize, you must select six numbers that are the same as the ones drawn randomly without replacement from a pool of, say, 40 numbers. The number of distinct 6 numbers chosen from 40 numbers is $\binom{40}{6}$ = 3,838,380. So any one choice of 6 numbers has a probability of $1/3,838,380 = 10^{-7} \times 2.605\ldots$. So your chances of winning the lotto game is, as everyone knows, incredibly small.

Let us assume that to win a share of second prize, you must select six numbers, five of which are amongst the six numbers drawn randomly without replacement from a pool of, say, 40 numbers. $\binom{6}{5}$ is the number of ways of choosing the 5 numbers from the 6 winning numbers. Then there are 6−5 numbers of the 6 numbers you chose which are from the 40 − 6 = 34 losing numbers, and there are $\binom{34}{6-5}$ ways of doing so. The probability of winning a share of second prize is

$$\frac{\binom{6}{5} \times \binom{34}{6-5}}{\binom{40}{6}} = 0.00005314\ldots.$$

More generally we can consider the following. Our lottery has N numbered balls and the player chooses n of these numbers. Let us see what the probability is that k of the players chosen numbers are amongst the K winning numbers drawn without replacement from the N numbers. The above argument shows the probability is

$$\frac{\binom{K}{k} \times \binom{N-K}{n-k}}{\binom{N}{n}}.$$

What we have described is quite simply the *hypergeometric distribution*.

Let us now consider a variation of lotto known internationally as *powerball*. For our discussion, let us assume that to share in the first prize, you must select six numbers that are the same as the ones drawn randomly without replacement from a pool of, say, 40 numbers, and you must also choose one extra number which is the same as the "powerball" which is a number drawn randomly from a different pool of, say, 20 numbers. As we saw previously the probability of choosing the six numbers correctly equals 1/3,838,380. The probability of choosing the powerball correctly is 1/20. So the probability of choosing the six numbers correctly and the powerball correctly equals $1/3{,}838{,}380 \times 1/20 = 10^{-8} \times 1.302\ldots.$

2.5 Voltaire Discovers a Flaw in French Lotto and Becomes a Millionaire

François-Marie Arouet (1694–1778), [23], was a prolific French writer, philosopher, writer, and advocate for freedom of speech and separation of church and state. Following his incarceration at the Bastille in 1718, he adopted the nom de plume M. de Voltaire by which he is well-known today. In 1729 the French government started running a lottery on bonds it owned, in an attempt to promote the purchase of these bonds. Only bond holders could buy tickets in this lottery, and the price was pegged to the value of the bond. Winners would get the face value of the bond as well as a half-million-livre jackpot. Voltaire and his mathematician friend Charles Marie de La Condamine (1701–1774) recognized a serious flaw in the design of this lottery.

If you owned a bond worth a very small amount, you could buy the lotto tickets extremely cheaply, yet your lotto ticket had just as much of a chance of winning as someone who owned a bond for 100,000 livres and had to buy their ticket for 100 livres. Voltaire, de la Condamine, and 11 compatriots formed a syndicate, and by June 1730, all had made a sizeable amount. By this process Voltaire himself amassed a profit the equivalent today of over US$12 million, which he then invested wisely and became a rich man. For a colourful version of this story, see [24]: "The Château de Ferney became Voltaire's home for the rest of his life, offering him a secure base from which to conduct his long and resolute campaigns against church and state. The house had been expensive to build and was expensive to run. Accordingly, he began to invest successfully in silk production and watch manufacture. Other monies were invested more conventionally, for example, in the provision of loans to asset-rich but cash-poor aristocrats who needed to fund their lifestyle".

2.6 Bingo

The history of the popular game of *bingo* begins in the sixteenth century with the Italian lottery *Il Gioco del Lotto d'Italia*. The game spread to France from Italy and was known as *Le Lotto*, played by the French aristocracy. The game is thought to have spread to England in the eighteenth century. Players mark off numbers on a ticket as they are randomly called out by a *caller* in order to achieve a winning combination. The similar *Tombola* was used in nineteenth century Germany as an educational tool to teach children multiplication tables and spelling and one can still find Tombola used in the context of a Montessori Educational game.

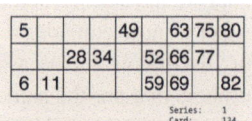

British Bingo Card

Housey-Housey cards

Bingo, previously known in the UK as *housey-housey*, became increasingly popular following the Betting and Gaming Act 1960 with purpose-built bingo halls. Bingo played in the UK (90-ball bingo) is slightly different from bingo played in the USA (75-ball bingo), as the tickets and the calling are slightly different.

2.6 Bingo

Problems

Dead Man's Hand in Poker

James Butler Hickok (1837–1876), well-known known as "Wild Bill" Hickok, [33], was a folk hero of the American Wild West and was a soldier, scout, lawman, gambler, and gunfighter. While playing poker in 1876, he was shot and killed by an unsuccessful gambler, who walked up behind him and shouted, "Damn you! Take that" before shooting him in the back of the head at point-blank range. The hand of cards which Hickok supposedly held at the time of his death has become known as the *dead man's hand*: ace of spades, ace of clubs, 8 of spades, and 8 of clubs (i.e. two black pairs), plus one other card.

Hickok

2.1 In a fair game of poker using a standard deck of 52 cards, what is the probability of being dealt a "dead man's hand"?

2.2 Assume you are dealt a "dead man's hand", where the fifth card is a 2 of spades. If you discard the 2 of spades and are dealt another card, now what is the probability that you will have a full house by having 3 aces or three 8s)?

2.3 Assume you are dealt a "dead man's hand", where the fifth card is a 5 of spades. If you discard all but the 2 aces and are dealt three other cards, what is the probability that you will have four aces?

2.4 Compare the probabilities of being dealt a hand with at least three kings in (a) poker, (b) bridge, (c) euchre, and (d) rummy.

2.5 In the game of rummy as described, calculate the probability of being dealt a hand with a sequence of length at least three.

2.6 In the game of rummy, calculate the probability of being dealt a hand which consists of both a four-card run and a three-card Set?

2.7 In a variant of the game of rummy described in there is a 54-card deck which is the standard 52 cards plus 2 jokers. Now each joker can behave like any card. So 5 of clubs-joker-7 of clubs is an acceptable run and queen of hearts-queen of spades-joker is an acceptable set. What is the probability in this version of rummy of being dealt (i) a four-card run or (ii) a four-card set?

2.8 There is a variant of lotto (not powerball). In this variant the player selects 6 numbers, but there are 8 numbers drawn without replacement from a pool of 40 numbers, with the last 2 of these drawn numbers called supplementary numbers. To win a share of first prize, the 6 numbers the player chooses need to be the same as the first 6 numbers of the 8. To win a share of second prize, 5 of the 6 numbers chosen by the player must be the same as 5 of the 6 numbers drawn, and the 6th number chosen by the player needs to be one of the two supplementary numbers. Check whether the probability of winning a share of second prize is greater than 10^{-7}.

Dice Problem of Samuel Pepys

Shortly we shall mention a probability problem that Samuel Pepys asked Sir Isaac Newton in a letter dated November 22, 1693.

Samuel Pepys F.R.S. (1633–1703) was a President of the Royal Society (the oldest continuously existing scientific academy in the world), an administrator of the Navy of England, and a Member of English Parliament, who is most famous for the diary he kept for a decade while still a young man. His diary, kept from 1660 until 1669, is one of the most important sources for eyewitness accounts of great events including the Great Plague of London, the Second Dutch War, and the Great Fire of London. Robert Latham (1912–1995), editor of the definitive edition of the diary, remarks concerning the Plague and Fire: "His descriptions of both–agonisingly vivid–achieve their effect by being something more than superlative reporting; they are written with compassion. As always with Pepys it is people, not literary effects, that matter".

Pepys

Diary volumes of Pepys

2.6 Bingo

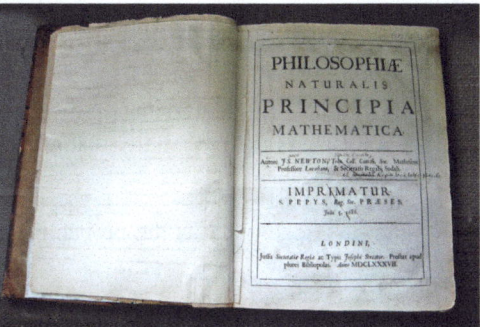

Newton — First edition of Principia mentioning Pepys

Sir Isaac Newton F.R.S. (1642–1726) was an English mathematician, physicist, theologian, astronomer, Master of the Royal Mint, a President of the Royal Society and a Member of Parliament, and one of the most influential scientists of all time. His book *Principia Mathematica* established classical mechanics. Newton also made contributions to optics and shares credit with Gottfried Wilhelm Leibniz (1646–1716) for developing calculus. He built the first reflecting telescope and described a theory of colour using the fact that a prism separates white light into the colours of the visible spectrum. He made the first theoretical calculation of the speed of sound, introduced the notion of a Newtonian fluid, contributed to the study of power series, and produced a method for approximating the roots of a function.

The letter referred to above from Pepys to Newton posed the following problem (which I have rephrased in modern language):

(A) Six fair dice are tossed and at least one ace appears. (The number 1 on a die is called an *ace*.
(B) Twelve fair dice are tossed and at least two aces appear.
(C) Eighteen fair dice are tossed and at least three aces appear.

[Pepys implicitly assumed the dice throws were independent.]

Which of the three has the biggest probability of success? Pepys thought that the correct answer is (C). Newton responded that the correct answer is (A) but gave no verification. When questioned further, Newton gave a simple logical argument which, according to [30], is faulty. Eventually Newton calculated the probabilities (at least for (A) and (B)) to show that (A) has the bigger probability and argued that by logic both of (A) and (B) have greater probability than (C). Newton's calculation is correct.

2.9 Verify that the probabilities of (A), (B), and (C) are, respectively, $0.6651\ldots$, $0.6186\ldots$, and $0.5973\ldots$.

[Hint for (A): Probability that six tossed fair dice produce at least one ace = 1− Probability that six tossed dice produce no aces.]

2.10 In the game of bridge, what is the probability that a player has either a 5-3-3-2 hand or a 5-4-3-1 hand?

2.11 In the game of bridge we showed that the probability that a player has a 4-4-3-2 hand is 0.2155.... What is the probability that player A and her partner both have 4-4-3-2 hands?

2.12 In the game of bridge, what is the probability that all 4 of the players have 4-4-3-2 hands?

Dreidels

The Jewish festival of Chanukah celebrates the success of the Maccabean revolt against the Helenistic King Antiochus IV Epiphanes about 2,200 years ago, capturing Jerusalem and rededicating the Second Temple. On the eighth day of festival, Jews light candles, eat foods fried in oil, and usually play with a *dreidel*. The dreidel is a four-sided spinning top, with a Hebrew letter on each side. Traditionally the letters were נ (nun), ג (gimmel), ה (hay), and ש (shin), which stands for the Hebrew words transliterated as nes gadol haya sham which means a great miracle happened there. Since the Jewish people returned to Israel, the Israeli dreidel has changed the letters on the dreidel to נ, ג, ה, פ (pey), to represent nes gadol haya po, a great miracle happened *here*.

The word Dreidel is in the Yiddish language and is derived from the verb dreyen meaning to turn. According to Jewish tradition, the dreidel dates back to the rule of King Antiochus IV Epiphanes when studying the Torah (Hebrew Bible) was illegal. The Jews studied the Torah and, when the Seleucids approached, hid the Torah and pretended to be engrossed in playing the dreidel game.

How is the dreidel game played? Each player puts one game piece, such as a penny, or a raisin, or chocolate gelt (chocolate money) into the centre pot. Then each player spins the dreidel once and sees which side is facing upwards after it stops. If נ faces up, the player gets nothing; if ג faces up, the player gets everything; if ה faces up, the player gets half of the pieces (rounding upwards if necessary) in the pot; if ש or פ faces up, the player puts one of their own pieces in the pot. If a player runs out of pieces, they exit the game. If at any stage the pot is empty, the game ends.

2.6 Bingo

The rules for the dreidel game are similar to the ancient four-sided *teetotum* game which typically have A, D, N, and T on the four sides. A teetotum is a form of gambling spinning top. It usually has four or six sides with a letter or number on each side. There are references to teetotum in the literature, for example, in the story by Lewis Carroll (the mathematician, Charles Lutwidge Dodgson (1832–1898) "Through the looking Glass", Alice provokes the White Queen to ask "Are you a child, or a teetotum". It is also referred to in Charles Dickens (1812–1870) "Our Mutual Friend" and in the 1845 work of Edgar Allan Poe (1809–1849), *The System of Doctor Tarr and Professor Fether*.

Dodgson   Dickens

Poe   Chardin

In the pictures below, we see an ancient Russian dreidel, a modern dreidel, a portion of the 1560 painting of Pieter Breugel the Elder (1526–1569) "Children's Games", and 1735 painting of Jean Baptiste-Siméon Chardin (1699–1779), now in the Louvre, "Boy with a Spinning Top or Child with a Teetotum". It is curious that in Volume XXV of the Journal of the Anthropological Institute of Great Britain and Ireland in 1896, there is an article By R. Etheridge Jr. called "The Game of Teetotum as practised by certain of the Queensland Aborigines" and a photo of teetotums of the (Australian) Aborigines.

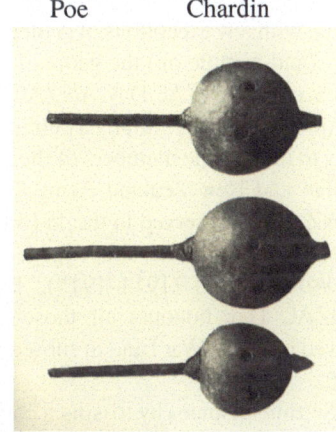

2.13 In the game of dreidels described above, there are four players, so that the game begins with four pieces in the pot. What is the probability that the game ends after exactly (i) one spin of the dreidel; (ii) two spins of the dreidel; (iii) three spins of the dreidel; (iv) four spins of the dreidel?

Two-Up

Two-up Pennies and Kip

In the photo above, we see two original 1915 Australian pennies in a kip (wooden bat) from which they are tossed in the game of *two-up*. In Australia, with few exceptions, it is illegal to play (and gamble on) the game of two-up except on ANZAC Day. ANZAC Day is April 25 and was initially a public holiday to honour the members of the Australian and New Zealand Army Corps (ANZAC) who served in the 1915 Gallipoli Campaign, their first engagement in World War I (1914–1918). Today ANZAC Day honours all those who served in World War I and in subsequent wars.

Australian soldiers playing two-up at the front, near Ypres in 1917

Settling disputes by tossing a coin has a history going back to the ancient Romans who called it *navia aut caput* which translates as ship or head as some coins had a ship on one side and the head of the emperor on the other side. In England this was referred to as *cross and pile*. Even today in cricket, tennis, and soccer, which side "starts" (bats, serves, or kicks off) is determined by the toss of a coin. The game of *two-up* was probably brought from England to Australia by convicts, as a variant of the *cross and pile*, a coin-tossing gambling game played with a single coin. (Between 1788 and 1868, about 162,000 convicts were forcibly transported from Britain and Ireland to penal colonies in Australia.) Two-up was played extensively by Australian soldiers during World War I. The game became a regular part of ANZAC Day activities throughout Australia by returned soldiers. Today as well as being played on ANZAC Day, it is played in some casinos throughout the year.

Using the kip, a player, known as the *spinner*, tosses the two pennies in the air. If both land with heads up, it is called as *heads*. If both land with tails up, it is called as *tails*. If one coin lands with a head up and the other with a tail up, it is called as *odds*.

2.6 Bingo 89

For our purposes, the spinner places a bet on heads and another player places an equal bet on tails. The rules are if the spinner tosses heads, they win the bet; if the spinner tosses tails, they lose the bet and the role of spinner; if the spinner tosses odds, they throw again except if the spinner throws odds five consecutive times, then the spinner and the other player betting lose their bets and the role of spinner. The money that has been bet goes to the *boxer*—the person managing the game.

2.14 In the game of two-up as described above, evaluate the probability

(i) that the spinner wins on their first toss;
(ii) that the spinner wins on their second toss;
(iii) that the spinner loses on their third toss;
(iv) that the boxer wins on the fifth toss;

2.15 In the game of mahjong, calculate the probability of having a 14-tile hand consisting of no seasons, no flowers, no pongs, no kongs, no cows, and no eyes.

2.16 In the game of mahjong, calculate the probability of having a 14-tile hand consisting of only honor tiles. (Observe that there are $7 \times 4 = 28$ honor tiles.)

2.17 In the game of mahjong, calculate the probability of a hand consisting of

(i) four chows plus eyes;
(ii) three pongs, one chow plus eyes;
(iii) two pongs, two chows plus eyes.

Birthday Problem

It is not entirely clear who first mentioned the birthday problem, but it is likely that it was Richard Edler von Mises (1883–1953) who was an Austrian Jewish scientist and mathematician who worked on solid mechanics, fluid mechanics, aerodynamics, aeronautics, statistics, and probability theory. The birthday problem is sometimes called the *birthday paradox*. The *birthday attack* is a type of cryptographic attack that exploits the mathematics behind this problem.

2.18 The *birthday problem* states that in a random group of 23 people, the probability that at least two people have the same birthday is greater than 0.5.

(i) Verify that this is true. (Recall that there are 365 or 366 days in each year.)
(ii) Verify that if the group has 100 people, rather than 23, the probability that 2 people in the group will have the same birthday is greater than 0.9999.

2.19 Find the probability of a 28 point hand in

(a) a two-person cribbage game;
(b) a three-person cribbage game.

2.7 Credit for Images

- Women's Whist Club Congress Drawing. Public Domain.
- Euchered lithograph 1884 from the Library of Congress. Public Domain.
- Joker. Public Domain.
- Playing Conquian in New Orleans. Public Domain
- Mah Jongg. Public Domain.
- Old Mahjong Tile Set displayed in the Tianyi Pavilion Museum. Licensed under the Creative Commons Attribution 3.0 Unported license. https://commons.wikimedia.org/wiki/File:Mahjong_museum.png
- Wild Bill Hickok. Public Domain.
- Old Mahjong Tile Set displayed in the Tianyi Pavilion Museum. Licensed under the Creative Commons Attribution 3.0 Unported license. https://commons.wikimedia.org/wiki/File:Mahjong_museum.png
- Andrey Andreyevich Markov. Public Domain.
- Teetotum: From "Children's Games" painted in 1560 by the Flemish artist Pieter Bruegel; Public Domain.
- Child With Teetotum. 1738 Painting in Louvre by Jean Siméon Chardin. Public Domain
- Two-up. This file is licensed under the Creative Commons Attribution-Share Alike 4.0 International license. https://commons.wikimedia.org/wiki/File:Two_up_set.jpg
- Two-Up game played by Australian soldiers. The photo is in the Australian War Memorial Museum. This image is of Australian origin and is now in the public domain because its term of copyright has expired. https://en.wikipedia.org/wiki/Two-up#/media/File:Two_Up.jpg
- Isaac Newton. Portrait by Godfrey Kneller, 1689. Public Domain.
- Wooden Dreidel. This work has been released into the public domain by its author, Roland Scheicher at German Wikipedia.
- Russian antique dreidel. This file is licensed under the Creative Commons Attribution-Share Alike 4.0 International license. https://commons.wikimedia.org/wiki/File:Russian_Dreidel_(1).jpg
- Pieter Brueghel's Children's Games (1560). Public Domain. https://en.wikipedia.org/wiki/Teetotum#/media/File:Teetotum_Bruegel.gif
- L'Enfant au toton—Jean Baptiste Siméon Chardin—Musée du Louvre Peintures RF 1705. Public Domain.
- Charles Lutwidge Dodgson. Public Domain.
- Conquian players. Public domain.
- Old Mahjong Tile Set displayed in the Tianyi Pavilion Museum. Licensed under the Creative Commons Attribution 3.0 Unported license. https://commons.wikimedia.org/wiki/File:Mahjong_museum.png
- Mahjong Tiles. Creative Commons Attribution-Share Alike 4.0 International license. https://commons.wikimedia.org/wiki/File:MJs9-.svg
- Hickok. Public Domain.
- . Photo of portrait of Pepys by John Hayls, 1666. Public Domain

- Six volumes of the diary manuscript of Pepys. Public Domain.
- Portrait of Isaac Newton 1689. Public Domain.
- Isaac Newton's personal copy of the first edition of his Principia Mathematica, bearing Pepys's name. This file is made available under the Creative Commons CC0 1.0 Universal Public Domain Dedication (1898). Public Domain.
- Portrait of Stuart Dodgson/ Lewis Carroll (1898). Public Domain.
- Portrait by Jeremiah Gurney (1867) of Charles Dickens. Public Domain.
- Poe (circa 1849). Public Domain.
- Self-portrait of Chardin in 1771. Public Domain.
- Two-Up Mint pennies. This file is licensed under the Creative Commons Attribution-Share Alike 4.0 International license. https://commons.wikimedia.org/wiki/File:Two_up_set.jpg
- Australian soldiers playing two-up during World War I at the front near Ypres, 23 December 1917. Public Domain.
- American Mahjong. This file is licensed under the Creative Commons Attribution-Share Alike 3.0 Unported license. https://commons.wikimedia.org/wiki/File:Majiang2.JPG
- Modern Cribbage Board. Permission is granted to copy, distribute and/or modify this document under the terms of the GNU Free Documentation License, Version 1.2. https://commons.wikimedia.org/wiki/File:120-hole_cribbage_board.jpg
- Housey-Housey. Public Domain. https://commons.wikimedia.org/wiki/File:HouseyHouseyCards.jpg
- British Bingo card. Public Domain.

References

1. Aubrey, J.: Brief Lives. The Boydell Press, Martlesham (1975)
2. Belcher, R.: Euchre for Beginners. Rupp Belcher (2021)
3. Benjamin, F.: Euchre Strategies. Lulu Press, Morrisville (2006)
4. Conway, R.S.: How To Play Euchre. Raymond S. Conway (2022)
5. Crisfield, D.W.: Bridge for Everyone: A Step by Step Guide to Rules, Bidding, and Play of the Hand. Morris Book Publishing LLC, Kearney (2010)
6. Davidson, E.: American Mah Jongg for Beginners: A Comprehensive and Easy-to-Follow Guide to Master the Game from Scratch,. Egle Davidson (2023)
7. Duke, A.: Vorhaus, Decide to Play Great Poker: A Strategy Guide to No-Limit Texas Hold'Em. Huntington Press, Las Vegas (2011)
8. Green, J.H.: Twelve Days In The Tombs: Or A Sketch Of The Last Eight Years Of The Reformed Gambler's Life. Kessinger Publishing, Whitefish (2007)
9. Heinz, A.: Mahjong A Chinese Game and the Making of Modern American Culture. Oxford University Press, Oxford (2021)
10. Hoyle, E.: A Short Treatise on the Game of Whist. Creative Media Partners, LLC, Sacramento (2018). First published (1742)
11. Jack, G.: Mahjong for Beginners: How to Play Chinese Mahjong for Absolute Beginners. Garcia Jack (2022)

12. James, J.: Poker Face: Mastering Body Language to Bluff, Read Tells and Win. Marlowe & Company, New York (2007)
13. Jerome, F., Dickson, S.: Poker Wit and Wisdom: Everything You'll Never Need to Know About Poker. Think Books, Dallas (2006)
14. Jones, H.: The Laws and Principles of Whist Stated and Explained, 18th edn. (Written Under Pseudonym of Cavendish, Thomas de La Rue, London, (1889) http://www.gutenberg.org/ebooks/51039)
15. Kistler, L.: Play Smarter and Win More Mahjong: Logic, Strategy, and Tactics. Larry Kistler (2023)
16. Klinger, R., Husband, P., Kambites, A.: Basic Bridge: The Guide to Good ACOL Bidding & Play. Cassell Group, London (1998)
17. Lawrence, M.: Insights on Bridge: Moments in Bidding, Book 1, Baron Barclay Bridge Supplies, Louisville (2019)
18. Lo, A.: The Book of Mah Jong: An Illustrated Guide.Tuttle Publishing, Tokyo (2001)
19. MacKinnon, R.F.: Bridge, Probability and Information. Master Point Press, Toronto (2010)
20. Marks, A.: Card Games Properly Explained: Poker, Canasta, Cribbage, Gin Rummy, Whist, and Much More. Skyhorse Publishing, New York City (2010)
21. McNeely, S.: Ultimate Book of Card Games: The Comprehensive Guide To More Than 350 Games. Chronicle Books LLC, San Francisco (2009)
22. Parlett, D.: The Penguin Book of Card Games. Penguin Books, London (1979)
23. Pearson, R.: Voltaire Almighty A Life in Pursuit of Freedom. BloomsburyPublishing, New York (2005)
24. Pearson, R.: Voltaire's Luck: The French Philosopher Outsmarts the Lottery, Lap[ham's Quarterly. https://www.laphamsquarterly.org/luck/voltaires-luck
25. Pascal, R.: Cribbage for Beginners. Reynold Pascal (2019)
26. Rep, J.: The Great Mahjong Book: History, Lore, and Play. Tuttle Publishing, Tokyo (2007)
27. Robson, A., Segal, O.: Partnership Bidding at Bridge: The Contested Auction. Faber and Faber, London (1994)
28. Sandberg, E.: A Beginner's Guide to American Mah Jongg. Tuttle Publishing, North Clarendon (2007)
29. Scarne, J.: Scarne's Encyclopedia of Card Games. Harper & Row Publishers, New York (1973)
30. Stigler, S.M.: Isaac Newton as a Probabilist. Statist. Sci. **21**, 400–403 (2006)
31. Tobin, T.: How Do I Play Rummy. Trev Tobin (2024)
32. Wenzel, J.: The Only Poker Book You'll Ever Need: Bet, Play, and Bluff Like a Pro. Adams Media, Avon Massachusetts (2004)
33. Wilstach, F.J.: Wild Bill Hickok: The Prince of Pistoleers. Doubleday, New York City (1926)
34. Wright, J.: (2001) The Mathematics Behind Euchre. An Honours Thesis. Ball State University, Muncle, Indiana. https://cardinalscholar.bsu.edu/bitstream/handle/handle/193211/W75_2001WrightJonathanJ.pdf?sequence=1
35. Zheng, W.Y., Walker, M., Blaszczynski, A.: Mahjong gambling in the Chinese-Australian community in Sydney: a prevalence study. J. Gambl. Stud. **26**, 441–454 (2010). https://doi.org/10.1007/s10899-009-9159-3

Chapter 3
Conditional Probability

Abstract

This third chapter on conditional probability begins with a dramatic example. We have a virus test which is extremely good in that if someone has the virus, then there is a 99% chance the test comes back positive and if the person does not have the virus, there is a 99% chance the test comes back negative.

Nevertheless when a population of one million people is tested where 1,000 people actually have the virus, if an individual tests positive, there is only a 10% chance he/she has the virus! So we test all the people who tested positive a second time, and then those who test positive a second time have a 90% chance of having the virus. Conditional probability helps us understand why this is so. We meet Kolmogorov's definition of conditional probability and the important Bayes theorem. We see the Bertrand box problem and the intriguing Monty Hall problem and the related three prisoners problem of Martin Gardner. If we have a 45% chance of living to age 85, what is our chance of living to 85 if we have already reached the age of 75? We introduce Bayesian theory and mention John Maynard Keynes "treatise on probability".

3.1 Conditional Probability

Intellectual Honesty

Our first example will be testing for a virus. This is very topical. And it is quite easy to discuss the probability issues. But in my opinion, it is intellectually dishonest to discuss virus testing when you have no idea what a virus really is and what a test might involve. Therefore, I have decided to describe in outline what a virus is and what testing involves. Of course this is done at a superficial level, but it is enough to get the flavour. The interested reader can easily follow up on these topics. For example, see [12–14, 17].

Viruses

A *virus* is the smallest type of parasite, ranging in size from 0.02μm to 1μm. (The *μm* is a unit of length known as a *micrometre* or a *micron* and is defined to be 1×10^{-6} m, i.e. one millionth of a metre or one thousandth of a millimetre or about 0.000039inches.) Viruses infect all life forms, from animals and plants to microorganisms, including bacteria. Indeed viruses can only replicate within the cells of animals, plants, and bacteria and therefore are regarded as parasites. A viral particle contains a nucleic acid (RNA or DNA) core surrounded by a protein coat and often enzymes. Viruses are classified according to whether the nucleic acid is single or double stranded, whether a viral envelope is present and how they replicate.

Louis Pasteur (1822–1895) was a French biologist, microbiologist, and chemist known for discoveries of the principles of vaccination and pasteurization and is remembered for his breakthroughs in the prevention of diseases. He created the first vaccines for rabies and anthrax. However, he was unable to find a causative agent for rabies and speculated about a pathogen too small to be detected by microscopes.

Pasteur

Charles Chamberland (1851–1908) was a French microbiologist who worked with Louis Pasteur. In 1884 Chamberland developed a type of filtration known today as the Chamberland filter or the Chamberland-Pasteur filter, a device with a filter that had pores that were smaller than bacteria, thus making it possible to pass a solution containing bacteria through the filter and having the bacteria completely removed from the solution.

Chamberland

3.1 Conditional Probability

 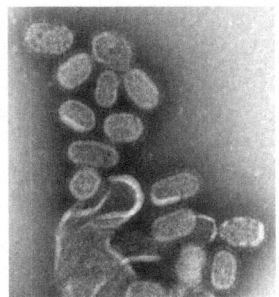 

Ivanovsky   influenza virus   Beijerinck

Dmitri Iosifovich Ivanovsky (1864–1920) was a Russian botanist, the discoverer of viruses, and one of the founders of the subject of virology. In 1892 he used the Chamberland-Pasteur filter to study what is now known as the tobacco mosaic virus: crushed leaf extracts from infected tobacco plants remained infectious even after filtration to remove bacteria. In 1898, the Dutch microbiologist and botanist Martinus Beijerinck (1851–1931) repeated Ivanovsky's experiments and became convinced that the filtered solution contained a new form of infectious agent. He observed that the agent multiplied only in cells that were dividing. He is credited with the conceptual discovery of viruses. Pictured above is the *influenza* virus (or *flu* virus) magnified 100,000 times. The tobacco mosaic virus was the first to be crystallized and its structure could, therefore, be elucidated. Rosalind Elsie Franklin (1920–1958), an English chemist and X-ray crystallographer, discovered the full structure of the virus in 1955 on the basis of her X-ray crystallographic pictures. Since then, thousands (of millions) of virus types have been described.

Cells, Chromosomes, and DNA

A *cell* is the basic building block of all living things. Your body is made of about 30 trillion = 3×10^{13} cells. Inside your body there are hundreds of different kinds of cells, each doing a different job. For example, red blood cells carry the oxygen you breathe around your body. *Chromosomes* are thread-like structures located inside the nucleus of cells. Each chromosome is made of protein and a single molecule of tightly wound DNA. So a chromosome is about 4 cm long, which is a very big molecule. Passed from parents to offspring, DNA contains the specific instructions that make each type of living creature unique. Tight packing allows the DNA to fit inside a tiny cell. There are 23 pairs of chromosomes in a cell. *Nucleic acids* are the main information-carrying molecules of the cell, and they determine the inherited characteristics of every living thing. The two main classes of nucleic acids are *DNA* and *RNA*. *Deoxyribonucleic acid* (DNA) is a molecule composed of two polynucleotide chains that coil around each other to form a double helix carrying genetic instructions for the development, functioning, growth, and reproduction of all known organisms and many viruses. DNA and *ribonucleic acid* (RNA) are nucleic acids. Both strands of double-stranded DNA store the same biological information. This information is replicated if and when the two strands separate.

The Swiss physician and biologist Johannes Friedrich Miescher (1844–1895) in 1869 discovered a microscopic substance in the pus of discarded surgical bandages and so became the first scientist to isolate nucleic acid. Rosalind Franklin (1920–1958) is best known for her work on the X-ray diffraction images of DNA, particularly Photo 51, which, unbeknown to her, led to the discovery of the DNA double helix for which Francis Crick and James Watson shared the Nobel Prize in 1962. Franklin was not eligible in 1962 as Nobel prizes are not awarded posthumously.

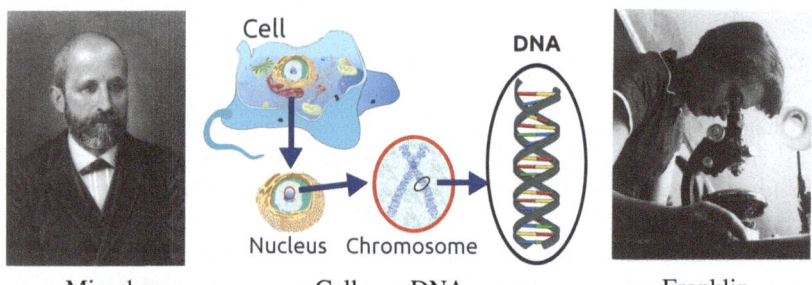

Miescher   Cell ⟶ DNA   Franklin

In the April 25, 1953, issue of the journal *Nature*, the Watson-Crick article, [15], was published proposing the double-helix structure of DNA. This was accompanied in the same issue by other articles providing evidence supporting it. They said about the proposed structure "It has not escaped our notice that the specific pairing we have postulated immediately suggests a possible copying mechanism for the genetic material".

James Dewey Watson (born April 6, 1928) is an American molecular biologist, geneticist, and zoologist. He shared with Francis Harry Compton Crick (1916–2004), a British molecular biologist, biophysicist, and neuroscientist, the Nobel Prize for the double-helix structure of the DNA molecule. Watson wrote an entertaining and informative book, [16] about this discovery. There is also a YouTube video, https://youtu.be/RvdxGDJogtA, of Watson describing the discovery and more. The role of Rosalind Elsie Franklin is discussed in Watson's book and in the article, https://www.nature.com/articles/nature01399, in the journal *Nature*.

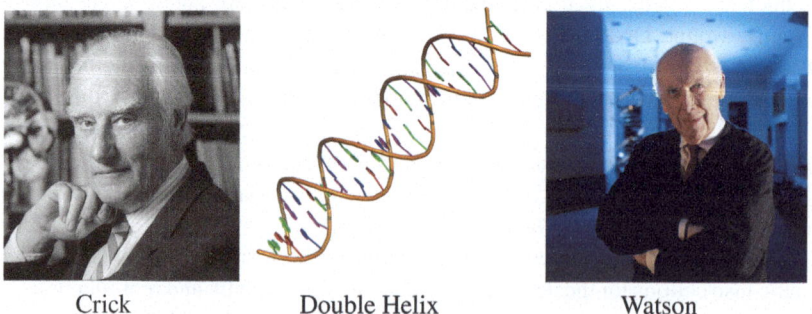

Crick   Double Helix   Watson

Polymerase Chain Reaction (PCR)

Kary Banks Mullis (1944–2019) was an American biochemist who won the Nobel Prize in Chemistry in 1983 for his invention of the *polymerase chain reaction (PCR)* technique. *The New York Times* described this technique as "highly original and significant, virtually dividing biology into the two epochs of before PCR and after PCR". His technique has become a central one in biochemistry and molecular biology.

Mullis

Enzymes are molecules (typically proteins) that significantly speed up the rate of chemical reactions that take place within cells. *Polymers* are a class of natural or synthetic substances composed of very large molecules that are multiples of simpler chemical units. A *polymerase* is an enzyme that synthesizes long chains of polymers or nucleic acids. Polymerase chain reaction (PCR) is a quick easy widely used method to make millions to billions of copies of a specific DNA sample within hours, allowing scientists to take a very small sample of DNA and amplify it to a large enough amount to study in detail. It is fundamental to much of genetic testing including identification of infectious agents. PCR is regarded as the "gold standard" in molecular diagnostics. The "gold standard" refers to the highest quality of a specific practice, product, or technology. PCR technology can be performed in hours to minutes compared to traditional methods such as culture, which is labour-intensive and can take days to produce a result.

In the case of COVID-19 testing, PCR can detect the genetic information (RNA) of the novel coronavirus, even if the virus is present in extremely small amounts.

Virus Testing

> *We now present an example which is extremely important in understanding conditional probability.*
> *Many people, even medical experts, jump to the wrong conclusion when dealing with conditional probabilities.*
> *We shall present other examples which demonstrate how badly our intuition fails.*

Example 3.1 Assume that we live in a country with a population of one million. To test for a dangerous virus, a *PCR* test is used. Assume also that a total of one thousand people in the country are infected with that virus.

There is a swab test for the virus and

(i) if someone tested has the virus, then the test comes back positive 99% of the time—this is called the *sensitivity* of the test. Also
(ii) if the person tested does not have the virus, then the test comes back negative 99% of the time—this is called the *specificity* of the test.

So it is a very effective test.

Now if Tom lives in that country and gets a test and it comes back positive, what is the probability he has the virus? Very high? You would think so. Let's analyse it.

(1) If all one million people are tested, then the one thousand people who are infected can expect 99% positive test results—that is, 990 positive test results. (And there are ten people who have the virus who test negative. These are called *false negative*.)
(2) Of the 999,000 who are not infected with the virus, 1% will test positive—that is, 9,990 will test positive. These are called *false positive*.

So altogether there are 9,990 + 990 = 10,980 positive test results. But 9,990 = over 90% of all positive test results are from people who do not have the virus. So astonishingly

if Tom gets a positive test result with this very effective test, there is nevertheless over 90% chance he does not have the virus.

(This is why, widespread testing of asymptomatic people is not a good idea.)

So what can we do? We thought we had a beautiful test, and it got it so wrong. The problem lies in misunderstanding conditional probability.

But is there a remedy for this virus testing? Yes!

Give a second test to everyone who has tested positive.

After the first test, we have 10,980 people who have tested positive, of whom 990 have the virus and 9,990 do not have the virus.

(a) When we test the 990 who have the virus, 99% will test positive, that is, 980 test positive. (Unfortunately, there are ten false-negative tests.)
(b) Of the 9,990 who do not have the virus, 1% will test positive, that is, about 100 test positive. These are false positives

So altogether 980 + 100 = 1,080 test positive a second time of whom 980 have the virus.

So if Tom tested positive the first time, was tested a second time, and again tested positive, the probability he is infected with the virus is $\frac{980}{1,080}$, that is, about 91% chance he has the virus.

So testing a second time all those with a positive first test improves the validity of the test result from 10 to 91%.

Remark 3.1 From the perspective of probability theory, we need to notice that there is a very great difference between the two events:

(i) let X be the event that a person tests positive, if they have the virus, and
(ii) let Y be the event that the person has the virus, if they test positive.

3.1 Conditional Probability

You may well ask how are the probability of event X occurring and the probability of event Y occurring are related. This will become clear in this section. For the time being, it suffices that you see the difference.

Remark 3.2 Example 3.1 applies not only to COVID-19 virus testing. It also applies to varying extents to other tests, for example, to mammograms (used to test for breast cancer), tests for HIV—human immunodeficiency virus—tests for influenza, troponin tests (for heart attacks), and serum myoglobin tests (for injury to the heart muscle).

Kolmogorov's Definition of Conditional Probability

Let us begin by recalling the definitions of σ-algebra, sample space, event space, and probability space and state a key proposition.

Definition 3.1 Let Ω be any set and Σ a set of subsets of Ω. Then Σ is said to be a *σ-algebra on* Ω if it satisfies the following four conditions:

(i) $\Omega \in \Sigma$;
(ii) if the subset S of Ω is in Σ, then its complement $S' = \Omega \setminus S \in \Sigma$;
(iii) if $S_1, S_2, \ldots, S_n \in \Sigma$, for some $n \in \mathbb{N}$, then $S_1 \cup S_2 \cup \cdots \cup S_n \in \Sigma$;
(iv) if $S_1, S_2, \ldots, S_n, \ldots$ are in Σ, then $\bigcup_{n=1}^{\infty} S_n \in \Sigma$.

Definition 3.2 Let Ω be a (finite or infinite) set, Σ a σ-algebra on the set Ω, and $I = \{1, 2, \ldots, n\}$ for some $n \in \mathbb{N}$ or $I = \mathbb{N}$. A *probability measure* or a *probability distribution*, P, is a function from Σ to the closed unit interval $[0, 1]$ with the following properties:

(i) $P(\emptyset) = 0$;
(ii) $P(\Omega) = 1$;
(iii) if sets $S_i \in \Sigma$, for $i \in I$, are such that $S_i \cap S_j = \emptyset$, for each $i, j \in I$ with $i \neq j$, then
$$P\left(\bigcup_{i \in I} S_i\right) = \sum_{i \in I} P(S_i).$$

In this context, the set Ω is said to be the *sample space* and the set Σ is said to be the *event space*. The triple (Ω, Σ, P) is said to be a *probability space*. If $A \in \Sigma$, then the number $P(A) \in [0, 1]$ is said to be the *probability of the event A* (or the *probability that A occurs*).

Proposition 3.1 *Let Ω be any sample space, Σ an event space, and (Ω, Σ, P) a probability space.*

(i) If $S \in \Sigma$, then $P(S') = P(\Omega \setminus S) = 1 - P(S)$.
(ii) If $A, B \in \Sigma$ and $A \subset B$, then $P(A) \leq P(B)$.
(iii) If $A, B \in \Sigma$, then $E = B \setminus (A \cap B) \in \Sigma$, the events A and E are mutually exclusive, and $A \cup B = A \cup E$ and hence $P(A \cup B) = P(A) + P(B) - P(AB)$.
(iv) Let $S_i \in \Sigma$, for $i \in \mathbb{N}$ or $i \in \{1, 2, \ldots, n\}$ for $n \in \mathbb{N}$. Then

$$P\left(\bigcup_{i \in I} S_i\right) \leq \sum_{i \in I} P(S_i),$$

with equality holding if the events S_i, $i \in I$, are mutually exclusive.

Proof. Exercise. □

Theorem 3.1 *Let (Ω, Σ, P) be a probability space and $A \in \Sigma$ an event such that $P(A) \neq 0$. The function $P_A : \Sigma \to [0, 1]$, defined by $P_A(B) = \dfrac{P(B \cap A)}{P(A)}$, for all $B \in \Sigma$, is a probability measure.*

Proof. We have to verify that P_A satisfies conditions (i), (ii), and (iii) of Definition 3.2. Firstly note that as Σ is a σ-algebra, $A, B \in \Sigma$ implies that $B \cap A \in \Sigma$, and so P_A does indeed map Σ into $\mathbb{R}$. By Proposition 3.1(ii), $P(B \cap A) \leq P(A)$ and so $P_A(B) = \dfrac{P(B \cap A)}{P(A)} \in [0, 1]$; that is, condition (i) of Definition 3.2 is indeed satisfied. Secondly observe that $P_A(\Omega) = \frac{P(\Omega \cap A)}{P(A)} = \frac{P(A)}{P(A)} = 1$, and so (ii) of Definition 3.2 is satisfied.
Thirdly, let sets $S_i \in \Sigma$, for $i \in I$, be such that $S_i \cap S_j = \emptyset$, for each $i, j \in I$ with $i \neq j$. Then as $(S_i \cap A) \cap (S_j \cap A) = \emptyset$, for $i \neq j$, using Definition 3.2 (iii) for P,

$$P_A\left(\bigcup_{i \in I} S_i\right) = \frac{P\left(\left(\bigcup_{i \in I} S_i\right) \cap A\right)}{P(A)} = \frac{P\left(\bigcup_{i \in I}(S_i \cap A)\right)}{P(A)} = \frac{\sum_{i \in I} P(S_i \cap A)}{P(A)}$$

$$= \sum_{i \in I} \frac{P(S_i \cap A)}{P(A)} = \sum_{i \in I} P_A(S_i), \text{ as required.}$$

□

We now state Kolmogorov's definition of conditional probability.

Definition 3.3 Let (Ω, Σ, P) be a probability space and A and B events in Σ such that $P(A) > 0$. The *conditional probability of B given A* (or *probability of B given A* is denoted $P(B \mid A)$ and equals $\dfrac{P(B \cap A)}{P(A)}$.

Remark 3.3 By Theorem 3.1 conditional probability is indeed a probability measure.

Bayes' Theorem

The most important result on conditional probability is known as *Bayes' Theorem*. It is named after the English statistician, philosopher, and Presbyterian minister Thomas Bayes (1701–1761) who formulated a specific case of the theorem.

Theorem 3.2 [Bayes' Theorem] *Let (Ω, Σ, P) be a probability space and A and B events in Σ such that $P(A) > 0$ and $P(B) > 0$. Then*

$$P(A \mid B) = \frac{P(B \mid A)P(A)}{P(B)}.$$

Proof. By Definition 3.3, $P(B|A) = \frac{P(B \cap A)}{P(A)}$ and so $P(B \cap A) = P(B \mid A)P(A)$. Similarly $P(A \cap B) = P(A \mid B)P(B)$. So $P(\mid A)P(A) = P(A \mid B)P(B)$. This proves the Theorem. □

Example 3.2 Let us look again at Example 3.1, but this time we shall use the terminology of conditional probability. Let $\Omega = \{A, B, C, D\}$, where
A = the person has the virus, A' = the person does not have the virus,
C = the person test positive, C' = the person tests negative.

Now Σ is the power set $\mathcal{P}(\Omega)$ which consists of all 16 subsets of Ω. We note that we are told that none of $P(A), P(A'), P(C), P(C')$ equal zero (since some people have the virus and some don't and some test positive and others don't). So we can use conditional probabilities as in Definition 3.3. We are given that

   (i) if the person has the virus, then they test positive; that is, $P(C \mid A) = 0.99$;
  (ii) if the person does not have the virus, then they test negative; that is, $P(C' \mid A') = 0.99$;
 (iii) 1,000 people out of 1,000,000 people have the virus; that is, $P(A) = 0.001$.
 (iv) $P(A') = 1 - P(A) = 1 - 0.001 = 0.999$:
  (v) $P(C \mid A') = 1 - P(C'|A') = 1 - 0.99 = 0.01$;
  (v) $P(C) = P(C \mid A).P(A) + P(C \mid A').P(A') = (0.99 \times 0.001) + (0.01 \times 0.999) = 0.01098$

Now we want to know what is the probability that the person has the virus if they test positive, that is, we want to know $P(A \mid C)$.

By Theorem 3.2,

$$P(A \mid C) = \frac{P(\mid A) \times P(A)}{P(C)} = \frac{0.99 \times 0.001}{0.01098} = 0.09016\ldots.$$

Bertrand's Box Problem

Bertrand   Chebyshev   Dedekind

The French mathematician Joseph Louis François Bertrand (1822–1900), who was a member of the Paris Academy of Sciences and its permanent secretary for 26 years, worked in number theory, differential geometry, probability theory, economics, and thermodynamics. In 1845 he conjectured that there is at least one prime number between n and $2n - 2$, for every $3 < n \in \mathbb{N}$. This was proved to be true by Pafnuty Lvovich Chebyshev (1821–1894) who was a Russian mathematician and is considered to be the founding father of Russian mathematics. This theorem is now known as *Bertrand's postulate*. It provides a different proof that there are an infinite number of prime numbers than the well-known proof of Euclid.

In 1849 Bertrand was the first to define real numbers using what is now called a *Dedekind cut*. Julius Wilhelm Richard Dedekind (1831–1916) was a German mathematician who made important contributions to abstract algebra, the axiomatic foundation for the natural numbers, algebraic number theory, and the definition of the real numbers.

Bertrand's box problem, known in the literature as *Bertrand's Box Paradox*, is a problem first posed by Joseph Bertrand in his 1889 work [2]. It is referred to as a paradox as the seemingly obvious answer to the problem is false and historically took mathematicians quite some time to accept the correct answer.

The problem is easily stated: There are three boxes:

1. a box containing two gold coins,
2. a box containing two silver coins, and
3. a box containing one gold coin and one silver coin.

After choosing a box at random, a coin is drawn from that box, and it is a gold coin. The problem asks: what is the probability that the other coin in that box is a gold coin? It *seems* that the probability is $\frac{1}{2}$ since it is equally probable that the box is (i) the box with two gold coins or (ii) the box with one gold coin and one silver coin. (Obviously it cannot be the box with two silver coins.) The probability is *not* $\frac{1}{2}$ as the statement that the two boxes are equally likely is false. The correct answer we shall see is $\frac{2}{3}$.

Example 3.3 We shall consider Bertrand's box problem as described above. We shall label the box with two gold coins GG, the box with one gold coin and one silver coin

3.1 Conditional Probability

as GS, and the box with two silver coins as SS. Each gold coin is labelled G, and each silver coin is labelled S. Now we shall examine this problem carefully using probability measure notation and solve it easily (and correctly) using conditional probabilities.

Consider the probability space (Ω, Σ, P) where

(i) the sample space $\Omega = \{GG, GS, SS\} \times \{G, S\}$, with the interpretation that the first coordinate represents the box chosen at random and the second coordinate represents the coin drawn from the chosen box. So the sample space has six members but two of them have probability 0, namely, (GG,S) and (SS,G).
(ii) the event space is the power set $\mathcal{P}(\Omega)$ of $2^5 = 32$ subsets,

where GG is choosing the box with two gold coins, etc., G is choosing a gold coin, and S is choosing a silver coin, so that, in particular:

(a) $P(GG) = P(GS) = P(SS) = \frac{1}{3}$, and
(b) $P(G) = P(S) = \frac{1}{2}$, since there are 3 gold coins and 3 silver coins.

This problem asks us to evaluate $P(GG \mid G)$. We shall apply Theorem 3.2 which will require us to know $P(G \mid GG)$, which equals 1.

$$P(GG|G) = \frac{P(G|GG) \cdot P(GG)}{P(G)} = \frac{1 \cdot \frac{1}{3}}{\frac{1}{2}} = \frac{2}{3}.$$

This example clearly demonstrates how using conditional probabilities makes even this controversial problem easy.

Monty Hall Problem

We shall now consider a famous problem which is very similar to Bertrand's box problem. It is known as the *Monty Hall problem*.

The Monty Hall problem is based on the American television game show "Let's Make a Deal" and named after its original host, Monty Hall. Monty Hall (1921–2017) was born as Monte Halparin to Orthodox Jewish parents. Let's Make a Deal aired at various times and on various networks from 1968 until 1991. The Monty Hall problem was originally posed (and solved) in a letter by the American University of Berkeley statistician Steve Selvin (born 1941) to *The American Statistician* in 1975. (*The American Statistician* is a quarterly peer-reviewed scientific journal, established in 1947, covering statistics and published on behalf of the American Statistical Association.)

Parade is an American nationwide Sunday newspaper magazine, distributed in over 700 newspapers in the USA, and has a circulation of 32 million. Marilyn vos Savant (born in 1946), who was listed as having the highest recorded intelligence quotient (IQ) in the Guinness Book of Records, has a column in *Parade* called "Ask

Marilyn" where she solves puzzles and answers questions on various subjects. In 1990 a reader's letter quoted in her column:

"Suppose you're on a game show, and you're given the choice of three doors: Behind one door is a car; behind the others, goats. You pick a door, say No. 1, and the host, who knows what's behind the doors, opens another door, say No. 3, which has a goat. He then says to you, 'Do you want to pick door No. 2?' Is it to your advantage to switch your choice?"

Vos Savant's response was that the contestant should switch to the other door, as she would have a 2/3 chance of winning the car, while if she stayed with door 1, her initial choice, she would have only a 1/3 chance.

Let us be clear, the host knows which door the car is behind.

Many readers of her column refused to believe switching is beneficial, despite her explanation. After the problem appeared in *Parade*, about 10,000 readers, including a large number of academics, wrote to the magazine, most of them claiming vos Savant was wrong. These included letters from the Deputy Director of the Center for Defense Information and a Research Mathematical Statistician from the National Institutes of Health—which contended that she was entirely incompetent. There is no doubt that sexism and misogyny were ingredients in some of the criticisms. Even when given explanations, simulations, and formal mathematical proofs, many people still do not accept that switching is the best strategy.

Intuition tells us that there are two doors remaining, and they are equally likely to have the car hidden behind. But this problem is counterintuitive.

Example 3.4 Consider the probability space (Ω, Σ, P) where

(i) the sample space $\Omega = \{D_1, D_2, D_3\} \times \{H_2, H_3\}$, with the interpretation that the first coordinate represents the door chosen at random and the second coordinate represents the item behind the chosen door, where D_1 is door 1 has the car, D_2 door 2 has the car, D_3 is door 3 has the car, H_2 is host opens door 2, and H_3 is host opens door 3.
(ii) the event space is the power set $\mathcal{P}(\Omega)$ of $2^5 = 32$ subsets,

where, in particular, $P(D_1) = P(D_2) = P(D_3) = \frac{1}{3}$.

We note that if the car is behind door 1, then the host can open door 2 or door 3.

$$\text{So we see that} \quad P(D_1 \cap H_2) = \frac{1}{3} \times \frac{1}{2} = \frac{1}{6} = P(D_1 \cap H_3)$$

We also note that if the car is behind door 2, then the host must open door 3, and if the car is behind door 3, the host must open door 2.

$$\text{So we see that} \quad P(D_2 \cap H_3) = \frac{1}{3} \times 1 = \frac{1}{3} = P(D_3 \cap H_2).$$

In fact, we are told the host opened door D_3. Let us evaluate $P(H_3)$. We know

$$P(H_3) = P(H_3 \cap D_1) + P(H_3 \cap D_2) + P(H_3 \cap D_3) = \frac{1}{6} + \frac{1}{3} + 0 = \frac{1}{2}.$$

By Definition 3.3, $P(D_1 \mid H_3) = \dfrac{P(D_1 \cap H_3)}{P(H_3)} = \dfrac{\frac{1}{6}}{\frac{1}{2}} = \dfrac{1}{3}.$

And $P(D_2 \mid H_3) = \dfrac{P(D_2 \cap H_3)}{P(H_3)} = \dfrac{\frac{1}{3}}{\frac{1}{2}} = \dfrac{2}{3}.$

So if the contestant stays with door 1, then they have a probability of only 1/3 of winning the car, but if they change to door 2, they have a probability of 2/3 of winning the car. So Marilyn vos Savant was correct.

Life Expectancy

Example 3.5 In Australia 75% of males live to age 75 or more and 47% of males live to age 85 or more. If a male person has reached the age of 75, what is the probability that he will live to age 85?

Consider the probability space (Ω, Σ, P), where $\Omega = \{E_n : n \in \mathbb{N}\}$, where E_n is that a male lives to at least age n and Σ is the power set $\mathcal{P}(\Omega)$. $P(E_n)$ is the probability that a male lives to at least age n. We are asked to evaluate $P(E_{85} \mid E_{75})$. Now, by Theorem 3.2,

$$P(E_{85} \mid E_{75}) = \dfrac{P(E_{75} \mid E_{85}) \times P(E_{85})}{P(E_{75})} = \dfrac{1 \times 0.47}{0.75} = 0.6266\ldots$$

So we see that the probability of a male who has lived to 75 living to at least 85 has increased by over 37% compared with the probability at birth of living to 85.

Remark 3.4 These days when life span is increasing it is likely that we will live for a longer period in retirement. Estimating whether our savings are sufficient to live reasonably in retirement requires us to estimate to what age we might expect to live. To do so, it is necessary to use conditional probabilities. The probability we seek is the conditional probability of living to a certain age if we know that we have lived to retirement age.

Bayesian Theory

We have presented probability theory from the viewpoint of Kolmogorov's axioms. There is an alternative, the *Bayesian theory* approach. Discussing Bayesian theory—Bayesian probability and Bayesian inference—would require a very substantial detour. So I will say just a few words which in no way do justice to this serious topic. I refer the interested reader to [1, 4, 7–9].

To approach Bayesian theory, it would be best if you suddenly forgot everything you know or think you know about probability. Bayesian theory is about *uncertainty*.

To explain what I am hinting at, I will tell you an anecdote.

In the early 1980s, I was in Ohio to do research with a colleague in Cleveland and to present a paper at a conference in Toledo. En route several colleagues and I visited an observatory with a sizeable telescope. We went into the building so as to look through the telescope. The first thing that happened was that our guide opened the two halves of the roof of the building. This seemed quite natural as we wanted to look at the sky. Then the walls of the building turned so the telescope could look in the right direction. Then the floor of the building we were standing on was moved upwards. At this point I wondered what in fact was fixed—not the roof, not the walls, and not the floor! It was unnerving. Approaching Bayesian theory should give you that same feeling.

The Italian probabilist, statistician, and actuary Bruno de Finetti (1906–1985) would say *there is no such thing as probability and independent events is a meaningless concept.*

These statements cause you to gasp as you have no firm foundation on which to rest, just like me in the observatory.

I started my discussion in Example 3.1 on conditional probability and virus testing with a city of one million people in which 1,000 people had the virus. The truth is that we can never know how many people have the virus.

Rather we can take a sample of the people and see how many in the sample have the virus. So we may have had an estimate of how many people have the virus before we sampled and then we modify our estimate on the basis of sampling.

In Bayesian theory, the conditional probability notion in Definition 3.3 takes on a different meaning. $P(A)$ may be our a priori estimate of the proportion that have the virus, and $P(A \mid C)$ is our *posterior* estimate in light of the sampling event C. Everything is an estimate!

Hopefully, I have said enough to whet your appetite to read very much more on this topic. I conclude by saying that some statisticians totally oppose Bayesian theory; others think it is absolutely the right approach.

John Maynard Keynes and Probability

In the previous section, I hoped to have whet your appetite about Bayesian probability.

Most of us have heard of John Maynard Keynes (1883–1946) who, according to Wikipedia, was an "English economist and philosopher whose ideas fundamentally changed the theory and practice of macroeconomics and the economic policies of governments". But few know that he wrote a deep thesis [6] called *A Treatise on Probability*.

Keynes

3.1 Conditional Probability

It is not possible to describe here in a few words what he understood by probability but it was much more than a number and much more related to logic. As explained in 2021 by Professor Jochen Runde, Professor of Economics & Organisation at Cambridge Judge Business School a century after the publication of Keynes work, "Keynes was proposing a novel and highly distinctive theory of probability of his own. In terms of this theory, probability is interpreted as a measure of the strength of a partial logical relation, what Keynes called the probability relation, between a hypothesis and the available evidence relevant to that hypothesis. A key feature of the theory is that probability relations are not generally capable of numerical measurement". William Peden in [11] also in 2021 wrote "In A Treatise on Probability, John Maynard Keynes (1921) provided the first systematic, subtle and self-conscious theory of what philosophers now call 'logical probability'".

Problems

3.1 Consider Example 3.1. Assume Tom gets a positive first test. Evaluate the probability that Tom has the virus after that first test and the probability of him having the virus after a positive second test if, instead of 1,000 people in the country having the virus:

(i) 10,000 people in the country have the virus; or
(ii) 100,000 people in the country have the virus.

3.2 Prove Proposition 3.1.

3.3 In Australia 92% of females live to age 65 or more and 41% of females live to age 90 or more. If a female has reached the age of 65, what is the probability that she will live to age 90?

3.4 Consider the Monty Hall problem discussed above. As stated previously, the host knows which door has the car behind it and chose to open a door that does not, namely, door 3. How do the calculations change if the host does *not* know which door the car is behind, so he chooses randomly between door 2 and door 3, chooses door 3, and the car is not behind it?

Martin Gardner's Three Prisoners Problem

Martin Gardner (1914–2010) was an American popular mathematics and popular science writer, with a special interest in the writings of Lewis Carroll. His work *The Annotated Alice* sold over a million copies. "The Three Prisoners problem" appeared in 1959 in his column "Mathematical Games" in Scientific American.

Gardner

3.5 Three prisoners, A, B, and C are in separate cells and sentenced to death. The prison governor has selected one of them at random to be pardoned. The warden knows which one is pardoned, but is not allowed to tell. Prisoner A begs the warden to let him know the identity of one of the two who are going to be executed. "If B is to be pardoned, give me C's name. If C is to be pardoned, give me B's name. And if I'm to be pardoned, secretly flip a coin to decide whether to name B or C".

The warden tells A that B is to be executed. Prisoner A is pleased because he believes that his probability of surviving has gone up from 1/3 to 1/2, as it is now between him and C. Prisoner A secretly tells C the news, who reasons that A's chance of being pardoned is unchanged at 1/3, but he is pleased because his own chance has gone up to 2/3. Which prisoner is correct?

[Hint. This problem is equivalent to the Monty Hall problem.]

Many problems in probability theory can be easily converted into problems about urns containing different numbers of balls of a variety of colours. We can represent contagious diseases, vehicles, atoms, etc. as coloured balls in an urn.

One of the first to consider an urn problem was Jacob Bernoulli (1654–1705) in his book on combinatorics and probability called "Ars Conjectandi". The problem concerned an urn containing a number of different coloured pebbles. Having randomly drawn several pebbles from the urn, the task was to estimate the proportion of each coloured ball in the urn. Jacob Bernoulli was one of the mathematicians in the Bernoulli family and supported Gottfried Wilhelm Leibniz in the Leibniz-Newton calculus controversy.

Bernoulli

We are familiar with problems concerning urns containing red and green balls. The urn is shaken thoroughly, and then, without looking, you put in your hand and draw out a ball and note the colour of the ball. In examples of *selection with replacement*, the ball that was drawn is returned to the urn before another ball is drawn out. In examples of *selection without replacement*, the ball is not returned before the next ball is drawn.

Leibniz

3.1 Conditional Probability

However, these are but two of a large variety of urn problems. A good reference for such problems is [5]. The authors observe, in particular, that many important probability distributions arise naturally in this context including the binomial distribution, the hypergeometric distribution, the normal distribution, the Poisson distribution, and the gamma distribution.

A special kind of urn problem, now known as Pólya's Urn, was introduced in 1923 in [3]. However, the idea appeared around 1906 in the work of Markov. The primary reference for this topic is [10].

George Pólya (1887–1985) was a Hungarian mathematician who made significant contributions to combinatorics, number theory, and probability theory. Andrey Andreyevich Markov (1856–1922) was a Russian statistician, whose best known work is what has become known as Markov chains and Markov processes.

Pólya

The Pólya urn introduced in [3] was to model communicable diseases. The urn contained balls of two colours. At each step, we shake the urn thoroughly, and a ball is sampled randomly. Each ball is equally likely to be selected. Each time a ball was drawn, its color is noted, then it is returned to the urn, and another ball of the same colour is also put in the urn. This is, in a sense, the exact opposite of sampling without replacement. You might say that the "rich get richer". For example, the more green balls in the urn, the greater is the probability of drawing a green ball which causes an additional green ball to be put in the urn. In contrast with selection without replacement, in a Pólya urn the balls in the urn are never exhausted. One can ask many questions about the balls in a Pólya urn after drawing a ball n times.

Markov

Of course the Pólya urn described above can easily be generalized.

One extension is to have an urn containing balls of up to k different colours. A ball is drawn from the urn. The colour of the ball is observed. Then C balls of that colour are returned. C is a constant, and, in principle, it can be any integer, positive, negative, or zero. Then we could ask how many balls of each colour are in the urn after the nth step of withdrawing a ball. An extra level of complexity can be added by replacing the constant C with a quantity that varies. For example, each time a ball is drawn, we throw a die, and the number that appears determines how many balls are put in the urn.

3.6 Consider the following Pólya urn. Initially there are r red balls and g green balls in the urn. Each time a coloured ball is drawn, that ball is returned, and an additional ball of that colour is placed in the urn. Let B_n be the event that the nth ball drawn is green and let G_n be the number of green balls drawn after n draws.

(i) Prove, using mathematical induction that $P(B_n) = \dfrac{g}{g+r}$.
   [Hint. First verify this for $n = 1$. Next assume it is true for $n = k$. Use $P(B_{k+1}) = P(B_{k+1} \mid B_1).P(B_1) + P(B_{k+1} \mid B_1').P(B_1')$, where B_1' is the complement of B_1, that is, the first ball drawn is red.]

(ii) If $r = g = 1$, verify that $P(G_n = m) = \binom{n}{m}\left(\dfrac{1}{2}\right)^n$.

3.2 Credit for Images

- Louis Pasteur. Public Domain.
- Charles Chamberland. Public Domain.
- Dmitri Iosifovich Ivanovsky. Public Domain. This work is in the public domain in Russia according to article 1281 of Book IV of the Civil Code of the Russian Federation No. 230-FZ of December 18, 2006, articles 5 and 6 of Law No. 231-FZ of the Russian Federation of December 18, 2006 (the Implementation Act for Book IV of the Civil Code of the Russian Federation).
  https://en.wikipedia.org/wiki/File:Ivanovsky.jpg
- Influenza virus. Public Domain.
- Martinus Willem Beijerinck. This file is licensed under the Creative Commons Attribution 4.0 International license.
  https://commons.wikimedia.org/wiki/File:Martinus_Willem_Beijerinck.png
- Johannes Friedrich Miescher. Public Domain.
- Calcul Des Probabilités. Fair Use.
  https://www.amazon.com.au/s?k=Calcul+Des+Probabilites&i=stripbooks&ref=nb_sb_noss
- Pafnuty Lvovich Chebyshev. Public Domain.
- Francis Harry Compton Crick. This file was published in a Public Library of Science journal. Their website states that the content of all PLOS journals is published under the Creative Commons Attribution 4.0 license (or its previous version depending on the publication date), unless indicated otherwise.
  https://commons.wikimedia.org/wiki/File:Francis_Crick_crop.jpg
- DNA ⟶ DNA. his file is licensed under the Creative Commons Attribution-Share Alike 3.0 Unported license.
  https://commons.wikimedia.org/wiki/File:Eukaryote_DNA-en.svg
- Rosalind Elsie Franklin. This file is licensed under the Creative Commons Attribution-Share Alike 4.0 International license.
  https://commons.wikimedia.org/wiki/File:Rosalind_Franklin.jpg

- Double Helix: Double stranded DNA with coloured bases. This file is licensed under the Creative Commons Attribution-Share Alike 4.0 International license. https://commons.wikimedia.org/wikiFile:Double_stranded_DNA_with _coloured_bases.png
- James Dewey Watson. Public Domain.
- Kary Banks Mullis. Public Domain. https://commons.wikimedia.org/wiki/File:Kary_Mullis.jpg
- Joseph Louis François Bertrand. Public Domain.
- Wooden Dreidel. This work has been released into the public domain by its author, Roland Scheicher at German Wikipedia. https://commons.wikimedia.org/wiki/File:Dreidel_001.jpg
- Ancient Russian Enamel Zolotnik Dreidel. Creative Commons Attribution-Share Alike 4.0 International license https://commons.wikimedia.org/wiki/File:Russian_Dreidel_(1).jpg
- Euchered lithograph 1884 from the Library of Congress. Public Domain.
- Richard Dedekind. Public Domain
- Martin Gardner. This file is licensed under the Creative Commons Attribution-Share Alike 2.0 Germany license. https://en.wikipedia.org/wiki/Martin_Gardner#/media/File:Martin_Gardner .jpeg
- Jacob Bernoulli. Public Domain
- Ars Conjectandi. Public Domain.
- Gottfried Wilhelm (von) Leibniz. Public Domain.
- John Maynard Keynes. Public Domain.

References

1. Bernado, J.M., Smith, A.F.M.: Bayesian Theory. Wiley, Chichester (1994)
2. Bertrand, J.: Calcul des probabilités. Gauthier-Villars, Paris (1889)
3. Eggenberger, F., Polya, G.: Über die stastik verketteter vorgänge. Z. Angew. Math. Mech. **1**, 279–289 (1923)
4. De Finetti, B.: Theory of Probability—A Critical Introductory Treatment Treatment, Translated by Antoni Machi and Adrian Smith. Wiley, Hoboken (1979)
5. Johnson, J., Kotz, S.: Urn Models and Their Application: An Approach to Modern Discrete Probability Discrete Probability Theory. Wiley, New York (1977)
6. Keynes, J.M.: A Treatise on Probability. Macmillan, London (1921)
7. Lindley, D.: Introduction to Probability and Statistics from a Bayesian Viewpoint Part I: Probability, New Edition. Cambridge University Press, Cambridge (1980)
8. Lindley, D.: Introduction to Probability and Statistics from a Bayesian Point of View Part 2: Inference. Cambridge University Press, Cambridge (1965)
9. Lindley, D: Understanding Uncertainty. Wiley, Hoboken (2014)
10. Mahmoud, H.M.: Pólya Urn Models. Chapman & Hall/CRC, Boa Raton (2009)

11. Peden, W.: Probability and arguments: Keynes's legacy. Cambridge J. Econ. **45**(5), 933–950 (2021). https://doi.org/10.1093/cje/beab021
12. Pierce, B.A.: Genetics: A Conceptual Approach, 7th edn. Macmillan Science & Education, New York (2020)
13. Robinson, T.R., Spock, L.: Genetics for Dummies. For Dummies. Wiley, Hoboken, New Jersey, USA (2020)
14. Skwarecki, B: Genetics 101: From Chromosomes and the Double Helix to Cloning and DNA Tests, Everything You Need to Know about Genes. Adams Media, Avon (2018)
15. Watson, J.D., Crick, F.H.C.: A structure for deoxyribose nucleic acid. Nature **171**, 737–738 (1953). https://doi.org/10.1038/171737a0
16. Watson, J.D.: DNA: The Secret of Life. Alfred A. Knopf, New York City (2003)
17. Windelsprecht, M.: Genetics !01. Greenwood Press, Westport (2007)

Chapter 4
Stirling's Approximation Formula and Improvements

Abstract

Stirling's approximation formula for $n!$, the gamma function as an extension of the factorial to the complex numbers, improvements of Stirling's approximation up to the twenty-first century, the Gaussian integral, limits, sequences, series, infinite products, infinite integrals, Taylor series, the standard normal distribution, the gamma distribution, the exponential distribution, the chi-square distribution, evaluating zeta(2), male births, carefree couples, and the probability that two randomly chosen natural numbers are coprime.

4.1 Stirling's Approximation Formula

In Chap. 2 we saw the importance of permutations and combinations to probability theory. In both of these, we saw that $n!$ was central. But how big is $n!$ and how fast does it grow as n grows? We know that $\log n$ grows slowly compared with n and n grows slower than $n^2, n^3, \ldots$, and each of these grows slower than e^n. Where does $n!$ fit on this list? Since $n! = n.(n-1).(n-2).\ldots.2.1$, we see $n! < n^n$ and clearly as n grows, $n!$ gets bigger than $n^2, n^3, \ldots$.

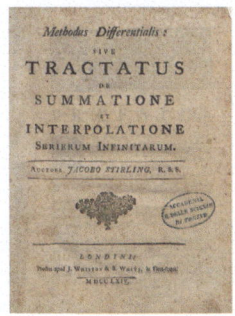

Stirling's approximation formula gives a good approximation of $n!$ when n is large, but, as we see in the next diagram and table, even when n is small. It was named after the Scottish mathematician James Stirling (1692–1770). The French mathematician Abraham de Moivre FRS (1667–1754) proved that $n!$ can be approximated by $cn^{n+\frac{1}{2}}e^{-n}$, for some constant c, where he gave an approximation for c. About 1730 Stirling showed that the constant c is $\sqrt{2\pi}$. De Moivre was analysing games of chance, and in that context, he needed a good approximation for $\binom{2n}{n}$ for large n. We saw in Sect. 1.3 that Cardano wrote the first book about games of chance. Another early book on this topic was "The Doctrine of Chances" and it was published in 1718 by de Moivre.

To understand Stirling's approximation, we need to discuss concepts which you may have met in first year calculus. However, I shall not assume you know this material, so do not be afraid. These topics are:

- convergence of sequences;
- infinite series; and
- infinite or improper integrals.

de Moivre

4.1 Stirling's Approximation Formula

We also meet the beautiful and very useful topic of Taylor series.

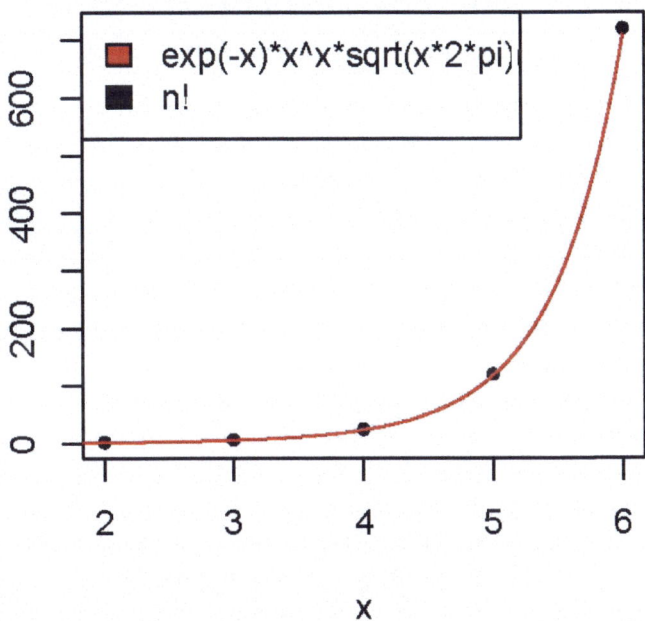

We plotted the above graph using the software package R and exported it to jpg format:

```
f1<-function(x) {
y<- (exp(-x)*x^x*sqrt(x*2*pi))
return(y)        }
value<-c(2,3,4,5,6)
  f2<-c(2,6,24,120,720)
  plot(value,f2, pch=20,  ylab="", xlab="x",
  main="Stirling's Approximation Formula")
  legend("topleft", legend=c("exp(-x)*x^x*sqrt(x*2*pi)",
  "n!"), ,   fill=c("red","black"))
  curve(f1(x),  xlim=c(0,6), ylim=c(0,1000), col="red",
   lwd=1, add=TRUE)
```

n	$n!$	Stirling's approx.	% error
2	2	1.919004	4.05
5	120	118.0192	1.65
10	3,628,800	3,598,696	0.83
20	2.432902×10^{18}	2.422787×10^{18}	0.42
50	3.041409×10^{64}	3.036345×10^{64}	0.17
100	9.332622×10^{157}	9.324848×10^{157}	0.08

We calculated the entries in the table above using the software package R:

```
f3<-function(x) {
y<- (100*(factorial(x)-f1(x)))/factorial(x)
return(y)       }
```

We see from the diagram and the table that for small n, the Stirling approximation is quite close to $n!$, and for larger n the approximation is very good.

4.2 Sequences and Limits

Now we introduce the notion of convergence of a sequence of numbers. The idea is quite simple: a sequence $x_1, x_2, \ldots, x_n, \ldots$ converges to a number a if as n gets bigger, x_n gets closer and closer to a. Indeed the x_n get arbitrarily close to a; for example, we can find an integer, $N_{0.1}$, such for all $n > N_{0.1}$, $|x_n - a| < 0.1$. We can also find an integer $N_{0.01}$, such for all $n > N_{0.01}$, $|x_n - a| < 0.01$. And we can also find an integer $N_{0.001}$, such for all $n > N_{0.001}$, $|x_n - a| < 0.001$, etc. Rather than keep writing smaller and smaller numbers 0,1, 0.01, 0.001,... we introduce a number ϵ which we can choose to be as small as we wish. This is expressed formally in our next definition.

Definition 4.1 If $x_1, x_2, \ldots, x_n, \ldots$ is a sequence of real numbers and $a \in \mathbb{R}$, then the sequence is said to *converge to a*, denoted by $x_n \to a$ or $\lim_{n \to \infty} x_n = a$, if for each positive real number ϵ, there exists an $N \in \mathbb{N}$ such that $|x_n - a| < \epsilon$ for all $n > N$. The sequence $x_1, x_2, \ldots, x_n, \ldots$ is said to be a *convergent sequence*.
A sequence which is not convergent is said to *diverge* or be a *divergent sequence*. (Some, but not all of the divergent sequences, will be *defined* to have their limit equal to ∞ or $-\infty$.)

4.2 Sequences and Limits

If for each positive integer M there is a positive integer N such that $x_n > M$, for all $n > N$, then we write $\lim_{n \to \infty} x_n = \infty$.

If for each positive integer M there is a positive integer N such that $x_n < -M$, for all $n > N$, then we write $\lim_{n \to \infty} x_n = -\infty$.

As an extension of the notion of the limit of a sequence introduced in Definition 4.1, we present:

Definition 4.2 Let $A \in \mathbb{R}$, S a subset of $\mathbb{R}$, and g a function from S to $\mathbb{R}$.

(i) If for each positive real number ϵ, there exists an $N \in \mathbb{R}$ such that for all $x > N$, $|g(x) - A| < \epsilon$, then $\lim_{x \to \infty} g(x)$ is defined to be equal to A.

(ii) If for each positive real number M there exists a positive integer N such that $g(x) > M$ for all $x > N$, then we write $\lim_{x \to \infty} g(x) = \infty$.

(iii) If for each positive real number M there exists a positive integer N such that $g(x) < -M$ for all $x > N$, then we write $\lim_{x \to \infty} g(x) = -\infty$.

(iv) If g is such that none of (i), (ii), and (iii) occurs, then $\lim_{x \to \infty} g(x)$ is said *not to exist*.

Example 4.1

(i) $1, \frac{1}{2}, \frac{1}{3}, \ldots, \frac{1}{n}, \cdots \to 0$.

(ii) $+1, -1, +1, -1, +1, -1, \ldots$ is a divergent sequence.

(iii) $+1, -\frac{1}{2}, +\frac{1}{3}, \ldots, (-1)^n \frac{1}{n}, \cdots \to 0$.

(iv) $\sin(1), \frac{\sin(2)}{2}, \ldots, \frac{\sin(n)}{n}, \cdots \to 0$.

(v) $1 \cdot e^{-1}, 2e^{-2}, \ldots, ne^{-n}, \cdots \to 0$. [Hint: $n < 2^n \implies ne^{-n} \leq \left(\frac{2}{e}\right)^n$, for $n \in \mathbb{N}$.]

(vi) $1 \cdot e^{-1}, 2^k e^{-2}, \ldots, n^k e^{-n}, \cdots \to 0$, for any $k \in \mathbb{N}$.

(vii) Let $f(x) = x \sin x$. Then $\lim_{x \to \infty} f(x)$ does not exist.

(viii) Let $f(x) = xe^{-x}$. Then $\lim_{x \to \infty} f(x) = 0$.

(ix) The sequence $1, \frac{1}{2}, 1, \frac{1}{3}, \ldots, 1, \frac{1}{n}, \ldots$ is divergent.

(x) If $a_n = -n$, for $n \in \mathbb{N}$, then $\lim_{x \to \infty} a_n = -\infty$.

(xi) Let $a_1, a_2, \ldots, a_n, \ldots$ be a sequence of real numbers with the property that $0 \leq a_{n+1} < a_n < 1$, for all $n \in \mathbb{N}$. Then $\lim_{n \to \infty} a_n$ exists. (This is left as an exercise.)

L'Hôpital's Rule

Limits are not always easy to determine by inspection. The following theorem provides a very useful tool in some cases. It is known as *L'Hôpital's rule*.

L'Hôpital's rule is named after the French mathematician Guillaume de L'Hôpital (1661–1704) as it first appeared in print in 1696 in the book he wrote which was the first textbook on differential calculus. However much of the material reported on in that book, including L'Hôpital's rule, came from material provided by the Swiss mathematician Johann Bernoulli (1667–1748)—in exchange for an annual payment of 300 Francs, Bernoulli would inform L'Hôpital of his latest mathematical discoveries, withholding them from correspondence with others. (L'Hôpital acknowledged his debt to Leibniz and the Bernoulli brothers, "especially the younger one" (Johann).)

Theorem 4.1 [L'Hôpital's Rule, [5, 27]]

(i) Let f and g be functions from an interval (a, b) to $\mathbb{R}$, where $a, b \in \mathbb{R}$ with $a < b$, such that the derivatives $f'(x)$ and $g'(x)$ exist for each $x \in (a, b)$. Further let $\lim_{x \to c} f(x)$ and $\lim_{x \to c} g(x)$ exist for some $c \in (a, b)$ and let both limits be equal to 0, and finally let $g'(x) \neq 0$ for each $x \in (a, b)$.
If $\lim_{x \to c} \frac{f'(x)}{g'(x)}$ exists and equals L, then $\lim_{x \to c} \frac{f(x)}{g(x)}$ exists and equals L.

(ii) Let f and g be functions from (a, ∞) to $\mathbb{R}$, where $a \in \mathbb{R}$ such that $f'(x)$ and $g'(x)$ exist for all $x \in (a, \infty)$ and $g'(x) \neq 0$ for all $x \in (a, \infty)$.
Let $\lim_{x \to \infty} f(x) = \lim_{x \to \infty} g(x) = 0$ or $\lim_{x \to \infty} f(x) = \lim_{x \to \infty} g(x) = \infty$ or $\lim_{x \to \infty} f(x) = \lim_{x \to \infty} g(x) = -\infty$.
If $\lim_{x \to \infty} \frac{f'(x)}{g'(x)} = L$, then $\lim_{x \to \infty} \frac{f(x)}{g(x)}$ exists and equals L.

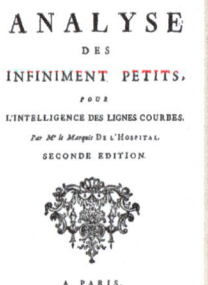

l'Hôpital's textbook

l'Hôpital

Johann Bernoulli

Example 4.2

(i) $\lim_{x \to 0} \frac{\sin x}{100x} = \lim_{x \to 0} \frac{\cos x}{100} = \frac{1}{100}$.

(ii) $\lim_{x \to 0} \frac{x}{\tan x} = \lim_{x \to 0} \frac{1}{\sec^2 x} = 1$.

(iii) $\lim_{x \to \infty} \frac{x}{e^x} = \lim_{x \to \infty} \frac{1}{e^x} = 0$.

Asymptotic

Definition 4.3 Let $x_1, x_2, \ldots, x_n, \ldots$ and $y_1, y_2, \ldots, y_n, \ldots$ be two sequences of real numbers. If all $y_n \neq 0$, and the sequence $\frac{x_1}{y_1}, \frac{x_2}{y_2}, \ldots, \frac{x_n}{y_n}, \ldots$ converges to 1, that is $\lim_{n \to \infty} \frac{x_n}{y_n} = 1$, then the two sequences are said to be *asymptotic* and this is denoted by $x_n \sim y_n$.

More generally, if f and g are functions from the interval (a, ∞) to $\mathbb{R}$, for $a \in \mathbb{R}$, such that $\lim_{x \to \infty} \frac{f(x)}{g(x)} = 1$, then $f(x)$ is said to be *asymptotic* to $g(x)$ and this is denoted by $f(x) \sim g(x)$.

Example 4.3 Let $x_n = n + \frac{1}{n}$ and let $y_n = n - \frac{1}{n}$, for all $n \in \mathbb{N}$. Both of the sequences $x_1, x_2, \ldots, x_n, \ldots$ and $y_1, y_2, \ldots, y_n, \ldots$ are divergent. However the sequence $\frac{x_1}{y_1}, \frac{x_2}{y_2}, \ldots, \frac{x_n}{y_n}, \ldots$ converges to 1. So $x_n \sim y_n$.

4.3 Series

Having discussed convergent sequences, now let us look at the notion of convergent series. Historically it was thought to be impossible that the sum of an infinite number of numbers could be finite. It was only in the seventeenth century that it was recognized that this can happen. But first one has to frame the problem correctly.

We know what a finite sum is:

$$S_n = \sum_{i=1}^{n} a_i = a_1 + a_2 + \cdots + a_n.$$

But what does $\sum_{i=1}^{\infty} a_i$ mean?. The answer is "nothing" yet. We have to give it a meaning, that is, we have to *define* it.

Definition 4.4 Let $a_1, a_2, \ldots, a_n \ldots$ be real numbers and define the number $S_n = \sum_{i=1}^{n} a_i = a_1 + a_2 + \cdots + a_n$. If $\lim_{n \to \infty} S_n$ exists and equals l, then the *sum of the infinite series* $\sum_{i=1}^{\infty} a_i$ is defined to be the number $l = \lim_{n \to \infty} S_n$. Such an infinite series is called a *convergent series*.

The series $\sum_{i=1}^{\infty} a_i$ is said to be *absolutely convergent* if the series $\sum_{i=1}^{\infty} |a_i|$ converges.

Example 4.4 It is not always easy to see immediately whether a series converges.

(i) The series $\sum_{i=1}^{\infty} \frac{1}{2^n} = \frac{1}{2} + \frac{1}{2^2} + \cdots + \frac{1}{2^n} + \ldots$ converges and equals 1. It is also absolutely convergent.

(ii) The series $\sum_{i=1}^{\infty} \frac{(-1)^n}{2^n} = -\frac{1}{2} + \frac{1}{2^2} + \cdots + \frac{(-1)^n}{2^n} + \ldots$ is clearly absolutely convergent.

(iii) The series $\sum_{i=1}^{\infty} \frac{1}{n} = 1 + \frac{1}{2} + \frac{1}{3} + \cdots + \frac{1}{n} + \ldots$ does not converge. This is far from obvious. The series is known as the *harmonic series*. That it diverges was first proved by the philosopher-mathematician Nicole Oresme ((about)1320–1382).

To see why this series diverges, we note the following:
$\frac{1}{3} > \frac{1}{4}$; $\frac{1}{5} > \frac{1}{8}$; $\frac{1}{6} > \frac{1}{8}$; $\frac{1}{7} > \frac{1}{8}$ etc.

So $1 + \frac{1}{2} + \left(\frac{1}{3} + \frac{1}{4}\right) + \left(\frac{1}{5} + \frac{1}{6} + \frac{1}{7} + \frac{1}{8}\right) + \cdots > 1 + \frac{1}{2} + \frac{1}{2} + \frac{1}{2} + \ldots$.

As the series on the right does not converge, clearly the harmonic series on the left does not converge.

Oresme

Leibniz

(iv) The series $\sum_{i=1}^{\infty} (-1)^n \frac{1}{n} = -1 + \frac{1}{2} - \frac{1}{3} + \cdots + (-1)^n \frac{1}{n} + \ldots$ converges but by (iii) is not absolutely convergent. (This follows from Leibniz's rule, [5, Theorem 10.14], which says that the alternating sequence $\sum_{n=1}^{\infty} (-1)^{n-1} a_n$ converges if (i) $a_{n+1} < a_n$, for all $n \in \mathbb{N}$ and (ii) $\lim_{n \to \infty} a_n = 0$. (Gottfried Wilhelm (von) Leibniz (1646–1716) was a prominent German polymath.)

4.4 Infinite Integrals

We now turn to *infinite integrals* (sometimes referred to as *improper integrals*). While we are familiar with integrals like $\int_a^b f(x)\,dx$ for any $a, b \in \mathbb{R}$ with $a < b$, we wish to assign a meaning to each of $\int_a^{\infty} f(x)\,dx$, $\int_{-\infty}^b f(x)\,dx$, and $\int_{-\infty}^{\infty} f(x)\,dx$. We need to do this now so we can introduce the Γ function.

4.4 Infinite Integrals

Definition 4.5 Let $a \in \mathbb{R}$ and f a function from (a, ∞) to $\mathbb{R}$. Let $g(x) = \int_a^x f(y)\,dy$. If $g(x)$ exists for all $x > a$ and $\lim_{x \to \infty} g(x)$ exists and equals $A \in \mathbb{R}$, then $\int_a^\infty f(y)\,dy$ is said to *converge* and is defined to be equal to A.

We now define the other two infinite integrals.

Definition 4.6 Let $a \in \mathbb{R}$ and f a function from $(-\infty, a)$ to $\mathbb{R}$. Let $g(x) = \int_{-x}^a f(y)\,dy$. If $g(x)$ exists for all $-x < a$ and $\lim_{x \to \infty} g(x)$ exists and equals $A \in \mathbb{R}$, then $\int_{-\infty}^a f(y)\,dy$ is said to *converge* and is defined to be equal to A.

Definition 4.7 Let f be a function from $(-\infty, \infty)$ to $\mathbb{R}$. If $\int_0^\infty f(x)\,dx$ and $\int_{-\infty}^0 f(x)\,dx$ converge, then $\int_{-\infty}^\infty f(y)\,dy$ is said to *converge* and is defined to be equal to $\int_{-\infty}^0 f(x)\,dx + \int_0^\infty f(x)\,dx$.

The following theorem is called the *Comparison Test for Infinite Integrals* and is [5, Theorem 10.24].

Theorem 4.2 *Let $a \in \mathbb{R}$ and let $g(x)$ be a function such that $\int_a^\infty g(x)\,dx$ converges. If $f : [a, \infty) \to \mathbb{R}$ is a function such that $\int_a^b f(x)\,dx$ exists for each $b \geq a$ and $0 \leq f(x) \leq g(x)$, for $x \in [a, \infty)$, then $\int_a^\infty f(x)\,dx$ converges and $\int_a^\infty f(x)\,dx \leq \int_a^\infty g(x)\,dx$.*
(The integral $\int_a^\infty g(x)\,dx$ is said to dominate the integral $\int_a^\infty f(x)\,dx$.)

Example 4.5

(i) We show here that $\int_1^\infty e^{-x^2}\,dx$ converges.

Putting $g(x) = e^{-x}$ we see that $\int e^{-x}\,dx = -e^{-x}$ and so $\int_1^\infty e^{-x}\,dx$ converges.

Put $f(x) = e^{-x^2}$. Then $1 \leq f(x) \leq g(x)$, for all $x \in [1, \infty)$. So by Theorem 4.2, $\int_1^\infty e^{-x^2}\,dx$ exists.

This shows the beauty of Theorem 4.2 in that we could show that $\int_1^\infty e^{-x^2}\,dx$ converges even though we couldn't evaluate easily the indefinite integral $\int e^{-x^2}\,dx$. We did it by comparing it with $\int e^{-x}\,dx$ which we could evaluate.

(ii) We can easily generalize (i) to show that, for any positive real number k, $\int_0^\infty e^{-kx^2}\, dx$ converges. To see this put $y = \sqrt{k}x$ so that $\frac{dy}{dx} = \sqrt{k}$ and hence $\int_0^\infty e^{-kx^2}\, dx = \frac{1}{\sqrt{k}} \int_0^\infty e^{-y^2}\, dy$. And we know that the right-hand side converges. So the left hand side exists too.

(iii) Noting that $\int_{-\infty}^\infty e^{-kx^2}\, dx = \int_0^\infty e^{-kx^2}\, dx + \int_{-\infty}^0 e^{-kx^2}\, dx = 2\int_0^\infty e^{-kx^2}\, dx$, which converges by (ii).

Trigonometry

Remark 4.1 At high school trigonometry was one of my favourite topics in mathematics. However, not everyone feels the same. So I will include a few pieces of basic information that we may need.

(i) $\tan x = \dfrac{\sin x}{\cos x}$;

(ii) $\sin^2 x + \cos^2 x = 1$;

(iii) $\sec x = \dfrac{1}{\cos x}$;

(iv) $\csc x = \dfrac{1}{\sin x}$;

(v) $\tan^2 x + 1 = \sec^2 x$;

(vi) $y = \sin x \iff x = \arcsin x$;

(vii) $y = \cos x \iff x = \arccos x$;

(viii) $y = \tan x \iff x = \arctan x$;

(ix) $\sin 0 = 0$, $\sin \frac{\pi}{4} = \frac{\sqrt{2}}{2}$, $\sin \frac{\pi}{2} = 1$, $-1 \leq \sin x \leq 1$, for all $x \in \mathbb{R}$;

(x) $\cos 0 = 1$, $\cos \frac{\pi}{4} = \frac{\sqrt{2}}{2}$, $\cos \frac{\pi}{2} = 0$, $-1 \leq \cos x \leq 1$, for all $x \in \mathbb{R}$;

(xi) $\tan 0 = 0$, $\tan \frac{\pi}{4} = 1$;

(xii) $y = e^x \iff \frac{dy}{dx} = y$ and $y = 0$ when $x = 1$;

(xiii) $\sinh x = \dfrac{e^x - e^{-x}}{2}$;

(xiv) $\cosh x = \dfrac{e^x + e^{-x}}{2}$;

(xv) $\tanh x = \dfrac{e^x - e^{-x}}{e^x + e^{-x}}$;

(xvi) $\dfrac{d(\sinh x)}{dx} = \cosh x$.

(xvii) $\dfrac{d(\cosh x)}{dx} = \sinh x$.

(xviii) $y = \ln x \iff x = e^y$. (Note that $\ln x$ is defined to be $\log_e x$.)

(xix) $\dfrac{d(\ln x)}{dx} = \dfrac{1}{x}$.

(xx) $\dfrac{d(x^n)}{dx} = nx^{n-1}$, for $n \in \mathbb{N}$.

(xxi) $\int x^n \, dx = \dfrac{x^{n+1}}{n+1} + C$, for $C \in \mathbb{R}$ and $n \in \mathbb{N}$.

(xxii) $\int \dfrac{1}{x} \, dx = \ln x + C$, for $C \in \mathbb{R}$.

4.5 Evaluating the Gaussian Integral $\int_{-\infty}^{\infty} e^{-x^2} \, dx$

Before proceeding to discuss another method to show certain infinite integrals exist, let us look further at Example 4.5 and actually evaluate the infinite integral $\int_{-\infty}^{\infty} e^{-x^2} \, dx$. This is known as the *Gaussian integral* and is named after the German mathematician Johann Carl Friedrich Gauss (1777–1855) who published this integral in 1809. It is also known as the *Euler-Poisson integral* after the Swiss polymath Leonhard Euler (1707–1783) and the French mathematician, engineer, and physicist Siméon Denis Poisson (1781–1840). The evaluation of the Gaussian integral below was by the French polymath Pierre-Simon, marquis de Laplace (1749–1827).

On the Eiffel Tower, 72 names of French scientists, engineers, and mathematicians are engraved in recognition of their contributions. Poisson is one of these.

Gauss

Euler

Poisson

There are many methods for evaluating the Gaussian integral. The most widely known proof is due to Poisson and uses polar co-ordinates and a double integral. The proof we present here, due to Laplace, also uses a different change of variable and depends only on single variable calculus.

Theorem 4.3 *Let $a, b \in \mathbb{R}$, $a > 0$. Then*

$$\int_{-\infty}^{\infty} e^{-a(x+b)^2} \, dx = \sqrt{\dfrac{\pi}{a}}.$$

In particular $\int_{-\infty}^{\infty} e^{-x^2} \, dx = \sqrt{\pi}$ *and* $\int_{-\infty}^{\infty} e^{-\pi x^2} \, dx = 1$.

Proof. It suffices to evaluate $\int_{-\infty}^{\infty} e^{-x^2} dx$, as the general result then follows from the change of variable $y = \sqrt{a}(x+b)$ and observing that $\frac{dx}{dy} = \frac{1}{\sqrt{a}}$.

Put $I = \int_0^{\infty} e^{-x^2} dx$, so that

$$I^2 = \left(\int_0^{\infty} e^{-x^2} dx\right)\left(\int_0^{\infty} e^{-y^2} dy\right) = \int_0^{\infty} \left(\int_0^{\infty} e^{-(x^2+y^2)} dx\right) dy.$$

Put $x = yt$, so that $\frac{dx}{dt} = y$. So

$$I^2 = \int_0^{\infty} \left(\int_0^{\infty} e^{-y^2(t^2+1)} y \, dt\right) dy = \int_0^{\infty} \left(\int_0^{\infty} y e^{-y^2(t^2+1)} dy\right) dt.$$

Noting that $\int_0^{\infty} y e^{-a^2} dy = \frac{1}{2b}$ for any $b > 0$, we see that

$$I^2 = \int_0^{\infty} \frac{1}{2(t^2+1)} dt = \lim_{s \to \infty} \int_0^s \frac{1}{2(t^2+1)} dt$$

$$= \frac{1}{2} \lim_{s \to \infty} \int_{t=0}^{t=s} \frac{1}{\tan^2 z + 1} \sec^2 z \, dz, \quad \text{where } t = \tan z \text{ and } \frac{dt}{dz} = \sec^2 z$$

$$= \frac{1}{2} \lim_{s \to \infty} \int_{t=0}^{t=s} \frac{1}{\sec^2 z} \sec^2 z \, dz = \frac{1}{2} \lim_{s \to \infty} \int_{t=0}^{t=s} 1 \, dz = \frac{1}{2} \lim_{s \to \infty} [\arctan t]_0^s$$

$$= \frac{1}{2}\left(\frac{\pi}{2}\right), \quad \text{as } \lim_{s \to \infty} \arctan s = \frac{\pi}{2}.$$

So $I = \int_0^{\infty} e^{-x^2} dx = \frac{\sqrt{\pi}}{2}$ and thus $\int_{-\infty}^{\infty} e^{-x^2} dx = \sqrt{\pi}$. $\square$

4.6 The Standard Normal Distribution

Remark 4.2 In Theorem 4.3 we proved that $\int_{-\infty}^{\infty} e^{-\pi x^2} dx = 1$. This fact is of considerable interest in statistics. It expresses the fact that the standard normal distribution is $\psi(x) = \frac{e^{-x^2}}{\sqrt{\pi}}$, that is, a *Gaussian Distribution* with mean $\mu = 0$ and *standard deviation* $\sigma = 1$. The graph of this distribution, usually referred to as *bell-shaped*, is below.

4.6 The Standard Normal Distribution

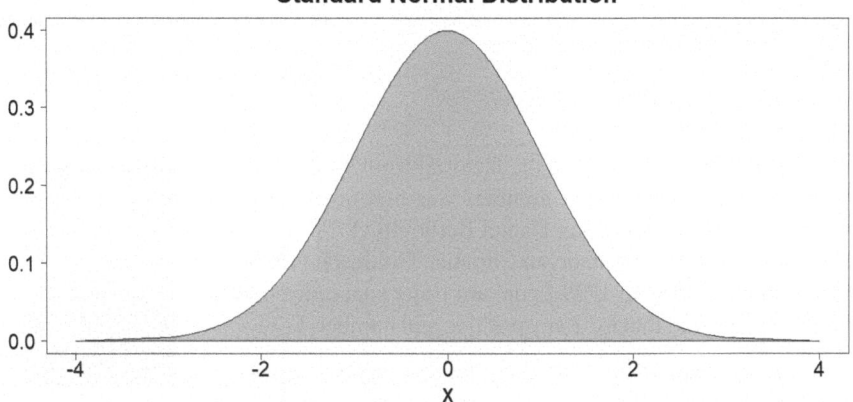

This graph was plotted using the software package R and exported to jpg format.

```
x=seq(-4,4,length=400)
y=dnorm(x)
plot(x,y,type="l",lwd=1, col="gray", cex.lab=2.0,
cex.axis=2.0, cex.main=2.5,
  cex.lab=2.5, ylab="", main="Standard Normal Distribution",
  las = 1)
x=seq(-3,3,length=400)
y=dnorm(x)
polygon(c(-4,x,4),c(0,y,0),col="gray")
```

It is interesting to note that the R software package can be used to obtain easily that the (shaded) area under the curve between $x = -3$ and $x = 3$ is $0.9973\ldots$.

```
pnorm(3,mean=0,sd=1)-pnorm(-3,mean=0,sd=1)
```

This result is statistics says that

virtually all numbers drawn from the standard normal distribution lie within three standard deviations of the mean.

Limit Comparison Test for Infinite Integrals

The following theorem is called the *Limit Comparison Test for Infinite Integrals* and is [5, Theorem 10.25].

Theorem 4.4 *Let $a \in \mathbb{R}$ and $f, g : [a, \infty) \to \mathbb{R}$ functions such that $f(x) \geq 0$ and $g(x) \geq 0$ for each $x \geq a$. Further, for each real number $b \geq a$, let $\int_a^b f(x)\, dx$*

and $\int_a^b g(x)\,dx$ both exist. If $\lim_{x\to\infty} \frac{f(x)}{g(x)} = c$, where $c \neq 0$, then $\int_a^\infty f(x)\,dx$ and $\int_a^\infty g(x)\,dx$ both converge or both diverge.

The problem of extending the factorial from the positive integers to a wider class of numbers was first investigated by the Swiss mathematician Daniel Bernoulli (1700–1782) and the German mathematician Christian Goldbach (1690–1764) in the 1720s. In 1729 Leonhard Euler succeeded and in 1730 he proved that for x any positive real number, $\Gamma(x) = \int_0^\infty t^{x-1} e^t\, dt$, where $\Gamma(n) = (n-1)!$, for any positive integer n. (This is true for complex numbers with positive real part.)

Daniel Bernoulli

The name gamma function is due to Adrien-Marie Legendre (1752–1833).

In 1774 Pierre-Simon Laplace noticed that Stirling's formula for $n!$ has a generalization to the gamma function, namely, that for x a positive real number, $\Gamma(x+1) \sim \sqrt{2\pi x}\left(\frac{x}{e}\right)^x$.

4.7 The Gamma Function

Definition 4.8 For y a positive real number, define

$$\Gamma(y) = \int_0^\infty x^{y-1} e^{-x}\, dx.$$

The function Γ is called the *gamma function*.

Proposition 4.1 *For every positive real number y, the gamma function $\Gamma(y) = \int_0^\infty x^{y-1} e^{-x}\, dx$ converges.*

Proof. In preparation for using the limit comparison test Theorem 4.4, let $f(x) = x^{y-1} e^{-x}$ and $g(x) = x^{-2}$. Noting that $\int_1^\infty x^{-2}\, dx$ converges and $\lim_{x\to\infty} \frac{f(x)}{g(x)} = \lim_{x\to\infty} \frac{x^{y-1} e^{-x}}{x^{-2}}$ converges, $\int_1^\infty x^{y-1} e^{-x}\, dx$ converges. Obviously $\int_0^1 x^{y-1} e^{-x}\, dx$ is finite. So $\Gamma(y) = \int_0^\infty x^{y-1} e^{-x}\, dx$ converges. □

4.7 The Gamma Function

Example 4.6 As another example of the use of the limit comparison test Theorem 4.4, let $f(x) = x^3 e^{-x^2/6}$ and $g(x) = x^{-3}$. Noting that $\int_1^\infty x^{-3}\, dx$ converges and $\lim_{x \to \infty} \frac{f(x)}{g(x)} = \lim_{x \to \infty} x^6 e^{-x^2/6}$ converges, $\int_1^\infty x^3 e^{-x^2/6}\, dx$ converges. So $\int_0^\infty x^3 e^{-x^2/6}\, dx$ converges.
[We shall use this result in the proof of Stirling's approximation formula.]

Integration by Parts

Remark 4.3 In 1715, the mathematician Brook Taylor (1685–1731) whom we shall discuss shortly, published a very useful integration "trick" familiar to first year calculus students called *integration by parts*. It is an immediate consequence of the following result on differentiating products: If $f(x)$ and $g(x)$ are differentiable functions, that is, they have a derivative for all $x \in [a,b]$, for $a, b \in \mathbb{R}$ with $a \le b$, then

$$\frac{d(f(x)g(x))}{dx} = \frac{d(f(x))}{dx} g(x) + f(x) \frac{d(g(x))}{dx};$$

that is, $(f(x)g(x))' = f'(x)g(x) + f(x)g'(x)$.

The corresponding result for integrals, which is easily derived from this says, for $a, b \in \mathbb{R}$ with $a \le b$:

$$\int_a^b f(x) g'(x)\, dx = [f(b)g(b) - f(a)g(a)] - \int_a^b f'(x) g(x)\, dx.$$

Example 4.7 Using integration by parts with $f(x) = x^y$ and $g'(x) = e^{-x}\, dx$, we obtain

$$\int_0^b x^y e^{-x}\, dx = [-b^y e^{-b} + 0^y e^0] + \int_0^b y x^{y-1} e^{-x}$$

$$= -b^y e^{-b} + y \int_0^b x^{y-1} e^{-x}\, dx$$

If we let $b \to \infty$, we obtain $\Gamma(y+1) = y\, \Gamma(y)$.

Example 4.8 Let us use integration by parts to evaluate the integral we showed converged in Example 4.6.

$$I = \int_0^\infty x^3 e^{-x^2/6}\, dx = \lim_{b \to \infty} \int_0^b x^3 e^{-x^2/6}\, dx.$$

Put $z = x^2/6$, so that $dz = \frac{x}{3}\, dx$. So

$$I_b = \int_0^b x^3 e^{-x^2/6}\, dx = \int_{x=0}^{x=b} (6z)\, x\, e^{-z} \frac{3}{x}\, dz = 18 \int_{z=0}^{z=b^2/6} z\, e^{-z}\, dz$$

$$= 18\left(\left[-\frac{b^2}{6} e^{-b^2/6} - 0\right] - \int_0^{b^2/6} (-e^{-z}\, dz)\right) \quad \text{by } Remark\ 4.3$$

$$= 18\left(-\frac{b^2}{6} e^{-b^2/6} - e^{-b^2/6} + 1\right)$$

Thus $\quad I = \lim_{b \to \infty} I_b = 18$.

Proposition 4.2
(i) $\Gamma(1) = 1$;
(ii) $\Gamma(y+1) = y\,\Gamma(y)$;
(iii) $\Gamma(n+1) = n\,\Gamma(n)$, for all $n \in \mathbb{N}$;
(iv) $\Gamma(n+1) = n!$, for all $n \in \mathbb{N}$.

Proof. Using Example 4.8 and mathematical induction, this result is easily proved. The details are left as an exercise. □

4.8 The Gamma Distribution

Remark 4.4 The gamma function leads us to mention the *gamma distribution*. It is a two-parameter family of continuous probability distributions. The two parameters are the *shape parameter* which we denote by k and the *scale parameter* which we denote by θ. It is defined by

For $k, \theta \in \mathbb{R}$ with $k, \theta > 0, \quad f(x) = \dfrac{\theta^k x^{k-1} e^{-\theta x}}{\Gamma(k)}, \quad$ where $x > 0, x \in \mathbb{R}$.

Erlang   Helmert   Pearson

The Danish mathematician and engineer Agner Krarup Erlang (1878–1929) invented the subjects of traffic engineering and *queueing theory* and in this context used the *Erlang Distribution*, which is the special case of the gamma distribution

4.8 The Gamma Distribution

for which the shape parameter k is a positive integer. The German mathematician Friedrich Robert Helmert (1843–1917) discovered the χ^2-*Distribution*, which is the special case of the gamma distribution with the scale parameter $\theta = 2$ in 1875/6. The English mathematician Karl Pearson (1857–1936) rediscovered the χ^2-Distribution in 1900. The χ^2 *test* is one of the most used techniques for hypothesis techniques. The special case of the gamma distribution with the shape parameter $k = 1$ is the *Exponential Distribution*.

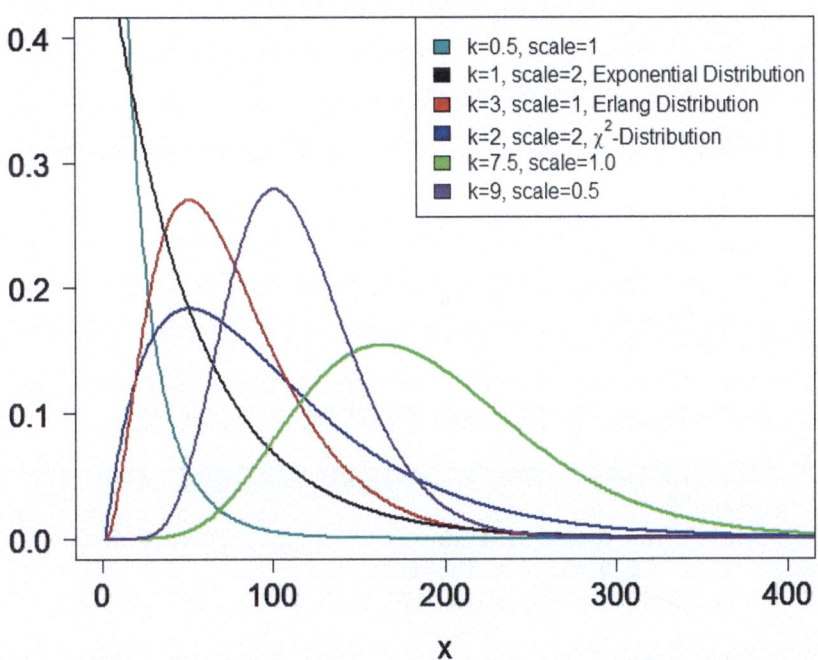

This graph was plotted using the software package R and exported to jpg format.

```
# R program to plot gamma distribution
# Specify x-values for gamma function
x_dgamma <- seq(0, 500, by = 0.04)
# Apply dgamma function

y_dgamma <- dgamma(x_dgamma, shape = 0.5, scale=1)

plot(y_dgamma, type="l", col="#009999", lwd=2, xlab="x",
ylab="",
  las=1,xlim=c(0,400), ylim=c(0,0.4), cex.lab=1.4,
  cex.axis=1.4,
```

```
    cex.main=1.6)
 par(new=TRUE)

  y_dgamma <- dgamma(x_dgamma, shape = 2, scale=2)

 plot(y_dgamma, type="l", col="blue", lwd=2, xlab="x",
  ylab="",
   las=1,xlim=c(0,400), ylim=c(0,0.4), cex.lab=1.4,
   cex.axis=1.4,
   cex.main=1.6)
 par(new=TRUE)

  y_dgamma <- dgamma(x_dgamma, shape = 3, scale=1)

 plot(y_dgamma, type="l",col="red", lwd=2, xlab="x",
  ylab="",
   las=1,xlim=c(0,400), ylim=c(0,0.4),
   cex.lab=1.4,
   cex.axis=1.4,
   cex.main=1.6)
 par(new=TRUE)

  y_dgamma <- dgamma(x_dgamma, shape = 1, scale=2)

  plot(y_dgamma, type="l", col="black", lwd=2, xlab="x",
  ylab="",
   main="The Gamma Distribution", las=1,xlim=c(0,400),
    ylim=c(0,0.4), cex.lab=1.4,
    cex.axis=1.4,cex.main=1.6)

 par(new=TRUE)

 y_dgamma <- dgamma(x_dgamma, shape = 7.5, scale=1)

 plot(y_dgamma, type="l", col="green", lwd=2, xlab="x",
 ylab="",
 main="The Gamma Distribution", las=1,xlim=c(0,400),
 ylim=c(0,0.4),
  cex.lab=1.4,
      cex.axis=1.4,cex.main=1.6)

 par(new=TRUE)

  y_dgamma <- dgamma(x_dgamma, shape = 9, scale=0.5)
```

4.8 The Gamma Distribution

```
    plot(y_dgamma, type="l", col="purple", lwd=2, xlab="x",
    ylab="",
     main="The Gamma Distribution", las=1,xlim=c(0,400),
      ylim=c(0,0.4),
      cex.lab=1.4,
      cex.axis=1.4,cex.main=1.6)

    x1 = expression(paste("k=2, scale=2, " , chi^2,
    "-Distribution"))

    legend("topright", legend=c("k=0.5, scale=1",
    "k=1, scale=2, Exponential Distribution",
    "k=3, scale=1, Erlang Distribution",
   x1,
    "k=7.5, scale=1.0",
     "k=9, scale=0.5 "),    fill=c("#009999", "black","red",
    blue", "green","purple"))
```

Taylor Series

We now discuss *Taylor Series*, introduced in 1715 by the English mathematician Brook Taylor. These series were used extensively by Colin Maclaurin (1698–1746) in the special case of the centre being zero.

While these series are very widely used in applied mathematics, physics, and engineering, many books and papers are a bit careless about their use. The fact is that the Taylor series of a function does not always converge, and even when it does converge, it does not necessarily converge to the function. (See [5, §11.13, Exercise 24].)

We state below Theorems 4.5, 4.6, and Proposition 4.3. These will suffice for most applications, and they are certainly adequate for our purposes.

Taylor

Maclaurin

Differentiable Functions

You will recall that a function which has a derivative is called *differentiable*. A functions which can be differentiated over and over again as many times as you like is called *infinitely differentiable* (or *smooth*). We know many infinitely differentiable functions, for example, all polynomials, and functions f such as given by $f(x) = \sin x$, $f(x) = \cos x$, and $f(x) = e^x$, and $f(x) = e^{-x}$. Of course the function f defined by $f(x) = |x|$ is not differentiable at $x = 0$.

Definition 4.9 Let $a \in \mathbb{R}$, r a positive real number and f a function from the open interval $(a-r, a+r)$ to $\mathbb{R}$. If f is infinitely differentiable on the interval $(a-r, a+r)$, then the *Taylor series generated by* f on the interval $(a - r, a + r)$ is defined to be

$$\sum_{k=0}^{\infty} \frac{f^{(k)}(a)}{k!}(x-a)^k.$$

If $a = 0$, then this series is said to be the *Maclaurin series generated by f*.

Theorem 4.5 [5, Theorem 11.11] *Let $a \in \mathbb{R}$, r a positive real number, and f a function from the open interval $(a-r, a+r)$ to $\mathbb{R}$. Assume that there exists a positive real number A such that $|f^{(n)}(x)| \leq A^n$, for all $n \in \mathbb{N}$ and every $x \in (a-r, a+r)$. Then, for every $x \in (a-r, a+r)$, the Taylor series generated by f equals $f(x)$.*

Another theorem of this type was proved by the Russian mathematician Sergei Natanovich Bernstein (1880–1968).

Unlike the previous theorem which requires the nth derivative $f^{(n)}$ not to grow too fast, here it suffices that all of its derivatives are non-negative on a closed interval $[0, r]$.

Bernstein

Theorem 4.6 [Bernstein's Theorem, [5, Theorem 11.12]] *Let r be a positive real number and f a function from $[0, r]$ to $\mathbb{R}$. If $f(x) \geq 0$ and $f^{(n)}(x) \geq 0$, for all $n \in \mathbb{N}$ and $x \in [0, r]$, then the Taylor series $\sum_{k=0}^{\infty} \frac{f^{(k)}(0)}{k!} x^k$ converges and equals $f(x)$, for all $x \in [0, r]$.*

The following simple proposition is useful.

Proposition 4.3 *Let f_1 be a function from $\mathbb{R}$ to $\mathbb{R}$ and f_2 a function from $[0, \infty)$ to $\mathbb{R}$. If for each positive real number r, the Taylor series for f_1 equals f_1 for $x \in (a-r, a+r)$, then the Taylor series for f_1 equals f_1 for all $x \in \mathbb{R}$.*

Also if for each $r \in (0, \infty)$, the Maclaurin series for f_2 equals f_2 for $x \in [0, r]$, then the Maclaurin series for f_2 equals f_2 for all $x \in [0, \infty)$.

With a little care, Corollary 4.1 can be easily deduced from Theorem 4.5, Theorem 4.6, and Proposition 4.3.

Corollary 4.1 [2]

(i) $\sin(x) = x - \dfrac{x^3}{3!} + \dfrac{x^5}{5!} - \dfrac{x^7}{7!} + \cdots + (-1)^{n-1}\dfrac{x^{(2n-1)}}{(2n-1)!} + \ldots$, for $x \in \mathbb{R}$.

(ii) $\cos(x) = 1 - \dfrac{x^2}{2!} + \dfrac{x^4}{4!} - \dfrac{x^6}{6!} + \cdots + (-1)^n\dfrac{x^{(2n)}}{(2n)!} + \ldots$, for $x \in \mathbb{R}$.

(iii) $e^x = 1 + x + \dfrac{x^2}{2!} + \cdots + \dfrac{x^n}{n!} + \ldots$ [$\implies e^x = \lim\limits_{x\to\infty}(1 + (1/x))^x$] for $x \in \mathbb{R}$.

(iv) $e^{-x} = 1 - x + \dfrac{x^2}{2!} + \cdots + \dfrac{(-1)^n}{n!} + \ldots$, & $e^{-x^2} = \sum\limits_{n=0}^{\infty}\dfrac{(-1)^n x^{2n}}{n!}$, $\forall x \in \mathbb{R}$.

(v) $\ln(1+x) = \sum\limits_{n=1}^{\infty}(-1)^{n+1}\dfrac{x^n}{n}$, for $|x| < 1$. [Recall $\ln$ is defined to be $\log_e$.]

(vi) $\dfrac{1}{1-x} = \sum\limits_{n=0}^{\infty} x^n$, for $|x| < 1$.

Proof.

$f(x) = \sin(x) \implies f'(x) = \cos(x) \implies f^2(x) = -\sin(x) \implies f^{(3)}(x) = -\cos(x)$
$\implies f^{(4)}(x) = \sin(x) \implies f^{(5)}(x) = \cos(x) \implies f^6(x) = -\sin(x)\ldots$

So $f^{2n-1}(x) = (-1)^{n-1}\cos(x)$ and $f^{(2n)}(x) = (-1)^n \sin(x)$.

Thus $f^{(2n)}(0) = 0$ and $f^{2n-1}(0) = (-1)^{(n-1)}$. Observe that $|\sin(x)| \le 1$ and $|\cos(x)| \le 1$ for all $x \in \mathbb{R}$.

The remainder of the proof is left as an exercise. □

4.9 Proofs from The Book

There are often many proofs of the same theorem. Some proofs are described as simple proofs, others as informative proofs, and yet others as elementary proofs. An informative proof is one which helps you understand the theorem. A simple proof is one which is usually short, but depends on some other results, sometimes powerful results, which you may or may not know. An elementary proof is one which uses the minimum background knowledge. The Hungarian Jewish mathematician Paul Erdös (1913–1996) described perfect proofs as "Proofs from The Book" [4] discusses such proofs.

Erdös

Plaque to the Stirlings of Garden, Dunblane Cathedral

4.10 Laplace Extension of Stirling's Formula

Theorem 4.7 [Laplace's Extension of Stirling's Formula]
*For x a positive real number, $\Gamma(x+1) \sim (x^x/e^x)\sqrt{2\pi x}$;
that is, $\displaystyle\lim_{x \to \infty} \frac{\Gamma(x+1)}{(x^x/e^x)\sqrt{2\pi x}} = 1$.
In particular, for n a positive integer, $n! \sim (n^n/e^n)\sqrt{2\pi n}$;
that is, $\displaystyle\lim_{n \to \infty} \frac{n!}{(n^n/e^n)\sqrt{2\pi n}} = 1$.*

Proof. Let $x, t \in \mathbb{R}, t, x > 0$. Further, let $f(x) = x^t e^{-x}$ and $A = \{x : |x-t| \geq \frac{t}{2}\}$. Let g_A be the characteristic function of A; that is, $g_A(x) = 1$, for $x \in A$ and $g_A(x) = 0$, otherwise. Then

$$\Gamma(t+1) = \int_0^\infty f(x)\,dx = \int_{\frac{t}{2}}^{\frac{3t}{2}} f(x)\,dx + \int_0^\infty g_A(x)f(x)\,dx \tag{4.10.1}$$

Observe that: $\quad x \in A \implies 1 \leq \dfrac{4(x-t)^2}{t^2};$ (4.10.2)

$$\Gamma(z+1) = z\Gamma(z), \text{ for all } z \in \mathbb{R}, z > 0: \tag{4.10.3}$$

4.10 Laplace Extension of Stirling's Formula

$$\Gamma(t+2) = (t+1)\Gamma(t+1); \tag{4.10.4}$$

$$\Gamma(t+3) = (t+2)(t+1)\Gamma(t+1). \tag{4.10.5}$$

$$\left| 1 - \frac{1}{\Gamma(t+1)} \int_{\frac{t}{2}}^{\frac{3t}{2}} x^t e^{-x}\, dx \right|$$

$$= \left| 1 - \frac{1}{\Gamma(t+1)} \left(\Gamma(t+1) - \int_0^\infty g_A(x) x^t e^{-x}\, dx \right) \right|$$

$$\leq \left| 1 - \frac{1}{\Gamma(t+1)} \left(\Gamma(t+1) - \int_0^\infty \frac{4(x-t)^2}{t^2} x^t e^{-x}\, dx \right) \right|$$

$$= \left| \frac{4}{t^2 \Gamma(t+1)} \int_0^\infty (x^2 + t^2 - 2xt) x^t e^{-x}\, dx \right|$$

$$= \left| \frac{4}{t^2 \Gamma(t+1)} \left(\int_0^\infty x^2 x^t e^{-x}\, dx + \int_0^\infty t^2 x^t e^{-x}\, dx - \int_0^\infty 2xt x^t e^{-x}\, dx \right) \right|$$

$$= \left| \frac{4}{t^2 \Gamma(t+1)} (\Gamma(t+3) + t^2 \Gamma(t+1) - 2t\Gamma(t+2)) \right|$$

$$= \left| \frac{4}{t^2 \Gamma(t+1)} ((t+2)(t+1)\Gamma(t+1) + t^2 \Gamma(t+1) - 2t(t+1)\Gamma(t+1)) \right|$$

$$= \frac{4}{t^2}(t^2 + 3t + 2 + t^2 - 2t^2 - 2t) = \frac{4}{t^2}(t+2).$$

Therefore
$$\lim_{t \to \infty} \frac{1}{\Gamma(t+1)} \int_{\frac{t}{2}}^{\frac{3t}{2}} x^t e^{-x}\, dx = 1. \tag{4.10.6}$$

Make the change of variables $x = y\sqrt{t} + t$ and define $h_t(y) = \left(1 + \frac{y}{\sqrt{t}}\right)^t e^{-y\sqrt{t}}$.

So
$$x = \frac{t}{2} \iff y = -\frac{\sqrt{t}}{2} \quad \text{and} \quad x = \frac{3t}{2} \iff y = \frac{\sqrt{t}}{2}. \tag{4.10.7}$$

Now
$$\int_{-\frac{\sqrt{t}}{2}}^{\frac{\sqrt{t}}{2}} h_t(y)\, dy = \int_{-\frac{\sqrt{t}}{2}}^{\frac{\sqrt{t}}{2}} \left(1 + \frac{y}{\sqrt{t}}\right)^t e^{-y\sqrt{t}}\, dy$$

$$= \int_{\frac{t}{2}}^{\frac{3t}{2}} \left(1 + \frac{x-t}{\sqrt{t}\sqrt{t}}\right)^t e^{-\left(\frac{x-t}{\sqrt{t}}\right)\sqrt{t}} \frac{1}{\sqrt{t}}\, dx = \frac{e^t}{t^t \sqrt{t}} \int_{\frac{t}{2}}^{\frac{3t}{2}} x^t e^{-x}\, dx.$$

By (4.10.6) and (4.10.7) this implies
$$\lim_{t\to\infty} \frac{t^t \sqrt{t}}{\Gamma(t+1)e^t} \int_{-\frac{\sqrt{t}}{2}}^{\frac{\sqrt{t}}{2}} h_t(y)\,dy = 1. \quad (4.10.8)$$

By Corollary 4.1 (v), $\ln(1+x) = \sum_{n=1}^{\infty}(-1)^{n+1}\frac{x^n}{n}$, for $|x| < \frac{1}{2}$. $\quad (4.10.9)$

So by (4.10.9) and Corollary 4.1(vi), for $z \in \mathbb{R}$ with $|z| < \frac{1}{2}$,

$$\left|\ln(1+z) - z + \frac{1}{2}z^2\right|$$

$$\leq \sum_{n=3}^{\infty}\frac{|z|^n}{n} \leq \sum_{n=3}^{\infty}\frac{|z|^n}{3} = \frac{|z|^3}{3}\sum_{n=3}^{\infty}|z|^{n-3} = \frac{|z|^3}{3}\frac{1}{1-|z|} \leq \frac{2}{3}|z|^3. \quad (4.10.10)$$

By Corollary 4.1 (iii), for all $u, v \in \mathbb{R}$

$$|e^u - e^v| = e^v|e^{u-v} - 1| = e^v\left|\sum_{n=1}^{\infty}\frac{(u-v)^n}{n!}\right|$$

$$\leq e^v|u-v|\sum_{n=1}^{\infty}\frac{|u-v|^{n-1}}{n!} \leq e^v|u-v|e^{|u-v|} \quad (4.10.11)$$

In (4.10.10) put $z = \frac{y}{\sqrt{t}}$ to obtain $\left|\ln\left(1 + \frac{y}{\sqrt{t}}\right) - \frac{y}{\sqrt{t}} + \frac{y^2}{2t}\right| \leq \frac{2|y|^3}{3t^{\frac{3}{2}}}. \quad (4.10.12)$

Put $\quad u = \ln h_t(y) = \ln\left(\left(1 + \frac{y}{\sqrt{t}}\right)^t e^{-y\sqrt{t}}\right) = t(\ln(1+z) - z). \quad (4.10.13)$

Put $v = -\frac{y^2}{2} = -\frac{tz^2}{2}$. Then for $|z| < \frac{1}{2} \iff |y| < \frac{\sqrt{t}}{2}$, we have

4.10 Laplace Extension of Stirling's Formula

$$\left|h_t(y) - e^{-\frac{y^2}{2}}\right| = \left|e^{t(\ln(1+z)-z)} - e^{-\frac{tz^2}{2}}\right|, \quad \text{by (4.10.13)}$$

$$\leq e^{-\frac{tz^2}{2}} \left|t(\ln(1+z) - z) + \frac{tz^2}{2}\right| e^{|t(\ln(1+z)-z)+\frac{tz^2}{2}|}, \quad \text{by (4.10.11)}$$

$$\leq e^{-\frac{y^2}{2}} \frac{t2|y|^3}{3t^{\frac{3}{2}}} e^{\left(\frac{2t|y|^3}{3t^{3/2}}\right)}, \quad \text{by (4.10.12)}$$

$$\leq \frac{|y|^3}{\sqrt{t}} \frac{2}{3} e^{y^2(-1/2 + \frac{2ty}{3t^{3/2}})}$$

$$< \frac{|y|^3}{\sqrt{t}} \frac{2}{3} e^{y^2(-1/2 + \frac{2t\sqrt{t}}{2 \cdot 3 \cdot t^{3/2}})}, \quad \text{as } y < \frac{\sqrt{t}}{2}$$

$$< \frac{|y|^3}{\sqrt{t}} e^{-\frac{y^2}{6}} \tag{4.10.14}$$

Therefore, by (4.10.14),

$$\left|\int_{-\frac{\sqrt{t}}{2}}^{\frac{\sqrt{t}}{2}} h_t(y)\,dy - \int_{-\infty}^{\infty} e^{-y^2/2}\,dy\right| \leq \frac{1}{\sqrt{t}} \int_{-\frac{\sqrt{t}}{2}}^{\frac{\sqrt{t}}{2}} |y|^3 e^{-y^2/6}\,dy + \int_{|y|>\sqrt{t}/2} e^{-y^2/2}\,dy$$

But the first integral on the right-hand side is finite by Example 4.8 and so the limit as $t \to \infty$ of the first term on the right-hand side is zero. The limit as $t \to \infty$ of the second integral is zero as $\int_{-\infty}^{\infty} e^{-y^2/2}\,dy$ is finite by Theorem 4.3. Therefore

$$\lim_{t \to \infty} \int_{-\frac{\sqrt{t}}{2}}^{\frac{\sqrt{t}}{2}} h_t(y)\,dy = \int_{-\infty}^{\infty} e^{-y^2/2}\,dy = \sqrt{2\pi}, \quad \text{by Theorem 4.3}$$

Combining this with (4.10.8), we have

$$\lim_{t \to \infty} \frac{\Gamma(t+1)e^t}{\sqrt{2\pi}t^{(t+1/2)}} = 1.$$

This completes the proof of the theorem, with the proof in fact yielding the result not only for $n!$, but in fact for the Γ function. □

4.11 Improvements on Stirling's Formula

Stirling's approximation of $n!$, proved about 1730, is surprisingly good and very useful outside statistics and pure mathematics.

There have been hundreds of published papers providing alternative proofs of Stirling's approximation or improving upon his approximation.

The English mathematician William Burnside (1852–1927) is known mostly as an early researcher in the theory of finite groups, but in his latter years, he turned to probability and wrote what was probably the first textbook in English on probability. It was published posthumously. As well, [7], published in 1917, gives a modest improvement on Stirling's approximation.

The first printed book on probability, published in 1657, was by the Dutch physicist, mathematician, astronomer, and inventor Christiaan Huygens (1629–1695). Huygens invented the pendulum clock and discovered Saturn's moon Titus. His book was directed towards games of chance.)

A significant improvement on Stirling's approximation and Burnside's approximation was made by the American mathematician and programmer Ralph William Gosper Jr. (born 1943) in 1978 in [11]. He also made significant contributions to computational mathematics, the MIT Maclisp system, and the powerful computer algebra package Macsyma.

A major advance was made by the extraordinary Indian mathematician Srinivasa Ramanujan (1887–1920) in the last year of his life. [24, p.339] made the claim that

$$\Gamma(x+1) = \sqrt{\pi} \left(\frac{x}{e}\right)^x \left(8x^3 + 4x^2 + x + \frac{\theta_x}{30}\right)^{\frac{1}{6}},$$

where $\theta_x \to 1$ as $x \to \infty$ and $\frac{3}{10} < \theta_x < 1$ and gave numerical evidence for it. Ramanujan's approximation was substantially better than those that preceded it and those which were discovered in the subsequent 80 years.

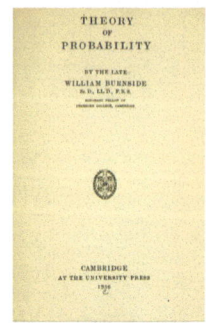

Burnside book

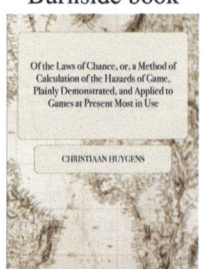

Huygens book

Huygens

Ramanujan

4.11 Improvements on Stirling's Formula

Since the year 2,000 there have been several papers extending Stirling's formula and improving on Gosper's result. In a web post in 2002, Robert H. Windschitl, [26], gave an elegant and good asymptotic approximation of $n!$, namely that

$$\Gamma(x+1) \sim \sqrt{2\pi x} \left(\frac{x}{e}\right)^x \left(x \sinh\left(\frac{1}{x}\right)\right)^{\frac{x}{2}}.$$

Burnside

In 2010 the Hungarian mathematician Gergő Nemes (born 1988) gave an asymptotic approximation which is almost as good as Windschitl's but better than all the others at that time. It was that

$$\Gamma(x+1) \sim \sqrt{2\pi x} \left(\frac{x}{e}\right)^x \left(1 + \frac{1}{12x^2 - \frac{1}{10}}\right)^x.$$

Gosper

An asymptotic formula of a different style, which is much better than Gosper's, was published in 2011 by the Romanian mathematician Cristinel Mortici, [17, 18]. It was

$$\Gamma(x+1) \sim \sqrt{2\pi x} \left(\frac{x}{e} + \frac{1}{12\,e\,x}\right)^x.$$

Pierre-Simon Laplace discovered what is now known as the *Stirling series* for the Gamma function.

$$\Gamma(x+1) \sim e^{-x} x^{x+\frac{1}{2}} \sqrt{2\pi} \left(1 + \frac{1}{12x} + \frac{1}{288x^2} - \frac{139}{51{,}840x^3} - \frac{571}{2{,}488{,}320x^4} + \sum_{n=5}^{\infty} \frac{a_n}{b_n x^n}\right),$$

where the real numbers a_n and b_n are explicitly calculated in [20]. As stated in [19], "the performance deteriorates as the number of terms is increased beyond a certain value".

In 2016 Chao-Ping Chen [9] produced an asymptotic approximation which was the best known at that time. It is

$$\Gamma(x+1) \sim \sqrt{2\pi x} \left(\frac{x}{e}\right)^x \left(1 + \frac{1}{12x^3 + \frac{24}{7}x - \frac{1}{2}}\right)^{x^2 + \frac{53}{210}}.$$

Ramanujan's approximation, as mentioned above, is substantially better than all those which were published in the subsequent 80 years. For example, when n equals one million, the percentage error of Ramanujan's approximation is one million million times better than Gosper's.

In 2013 the Australian mathematician Michael Hirschhorn (born 1947) and the Costa Rican mathematician Mark B. Villarino in [13] proved the correctness of Ramanujan's claim above for positive integers.

Hirschhorn

They showed that Ramanujan's θ_n satisfies for each positive integer n:

$$1 - \frac{11}{8n} + \frac{79}{112n^2} < \theta_n < 1 - \frac{11}{8n} + \frac{79}{112n^2} + \frac{20}{33n^3}.$$

Although they did not explicitly say it, it is clear from their work that

$$\Gamma(x+1) \sim \sqrt{\pi} \left(\frac{x}{e}\right)^x \left(8x^3 + 4x^2 + x + \frac{1 - \frac{11}{8x} + \frac{79}{112x^2}}{30}\right)^{\frac{1}{6}},$$

at least for positive integers.

In [16] I produced tables showing how good each of the approximations is for $n = 2$ to $n = 1$ million. Below we shall present just one table from which the flavour can be seen.

In [16] I observed that once we know that Stirling's formula is asymptotic to $\Gamma(x + 1)$, we can easily deduce that each of the other mentioned approximations also is asymptotic to $\Gamma(x + 1)$.

Theorem 4.8 *Let f be a function from an interval (a, ∞) to $\mathbb{R}$, where $a \in \mathbb{R}, a > 0$. If $\lim_{x \to \infty} f(x) = 1$, then $\Gamma(x + 1) \sim \sqrt{2\pi x} \left(\frac{x}{e}\right)^x . f(x)$.*

Proof. This is an obvious consequence of Laplace's extension of Stirling's formula Theorem 4.7. □

Example 4.9 $\Gamma(x + 1) \sim \sqrt{2\pi x} \left(\frac{x}{e}\right)^x . f(x)$, where $f(x) = 1 - 100x^{-7}$. However, while $\sqrt{2\pi x} \left(\frac{x}{e}\right)^x . (1 - \frac{100}{x^7})$ is asymptotic to $\Gamma(x + 1)$, it does not say it is a good approximation. How good each approximation is is demonstrated in the table below.

The following corollary follows immediately from Theorem 4.8. It shows that all of the asymptotic approximations we mentioned above are indeed asymptotic to $\Gamma(x + 1)$. This is interesting because it avoids a large amount of complicated

4.11 Improvements on Stirling's Formula

error estimate analysis and also because a couple of these approximations were only proved when x is a positive integer, whereas we get the results for all positive real numbers x.

Corollary 4.2 *For x a positive real number:*

(i) Burnside: $\Gamma(x+1) \sim \sqrt{2\pi} \left(\dfrac{x+1/2}{e} \right)^{x+1/2}$;

(ii) Gosper: $\Gamma(x+1) \sim \sqrt{\pi} \left(\dfrac{x}{e} \right)^x \sqrt{2x + \dfrac{1}{3}}$;

(iii) Mortici: $\Gamma(x+1) \sim \sqrt{2\pi x} \left(\dfrac{x}{e} + \dfrac{1}{12\,e\,x} \right)^x$;

(iv) Ramanujan: $\Gamma(x+1) \sim \sqrt{\pi} \left(\dfrac{x}{e} \right)^x \left(8x^3 + 4x^2 + x + \dfrac{1}{30} \right)^{\frac{1}{6}}$;

(v) Laplace (n): Fix $n \in \mathbb{N}$. For $a_i, b_i \in \mathbb{N}$,

$$\Gamma(x+1) \sim e^{-x} x^{x+\frac{1}{2}} \sqrt{2\pi} \left(1 + \frac{1}{12x} + \frac{1}{288x^2} + \sum_{i=3}^{n} \frac{a_i}{b_i x^i} \right);$$

(vi) Nemes: $\Gamma(x+1) \sim \sqrt{2\pi x} \left(\dfrac{x}{e} \right)^x \left(1 + \dfrac{1}{12x^2 - \frac{1}{10}} \right)^x$.

(vii) Windschitl: $\Gamma(x+1) \sim \sqrt{2\pi x} \left(\dfrac{x}{e} \right)^x \left(x \sinh \left(\dfrac{1}{x} \right) \right)^{\frac{x}{2}}$.

(viii) Hirschhorn and Villarino :

$$\Gamma(x+1) \sim \sqrt{\pi} \left(\frac{x}{e} \right)^x \left(8x^3 + 4x^2 + x + \frac{1 - \frac{11}{8x} + \frac{79}{112x^2}}{30} \right)^{\frac{1}{6}}.$$

(ix) Chen: $\Gamma(x+1) \sim \sqrt{2\pi x} \left(\dfrac{x}{e} \right)^x \left(1 + \dfrac{1}{12x^3 + \frac{24}{7}x - \frac{1}{2}} \right)^{x^2 + \frac{53}{210}}$.

Proof. In each case it is sufficient to determine the function f in Theorem 4.8 and observe that $\lim\limits_{x \to \infty} f(x) = 1$.

(i) Use $f(x) = \left(1 + \dfrac{1}{2x} \right)^x \left(\dfrac{1 + \dfrac{1}{2x}}{e} \right)^{\frac{1}{2}}$.

(ii) Use $f(x) = \sqrt{1 + \dfrac{1}{6x}}$.

(iii) Use $f(x) = \left(1 + \dfrac{1}{12x^2} \right)^x$.

(iv) Use $f(x) = \left(1 + \dfrac{1}{2x} + \dfrac{1}{8x^2} + \dfrac{1}{240x^3} \right)^{\frac{1}{6}}$.

(v) Use $f(x) = \left(1 + \dfrac{1}{12x} + \dfrac{1}{288x^2} + \sum_{i=3}^{n} \dfrac{a_i}{b_i x^i}\right)^x$.

(vi) Use $f(x) = \left(1 + \dfrac{1}{12x^2 - \frac{1}{10}}\right)^{\frac{x}{2}}$.

(vii) Use $f(x) = \left(x \sinh\left(\dfrac{1}{x}\right)\right)^{\frac{x}{2}}$.

(viii) Use $f(x) = \left(1 + \dfrac{1}{2x} + \dfrac{1}{8x^2} + \dfrac{1 - \frac{11}{8x} + \frac{79}{112x^2}}{240x^3}\right)^{\frac{1}{6}}$.

(ix) Use $f(x) = \left(1 + \dfrac{1}{12x^3 + \frac{24}{7}x - \frac{1}{2}}\right)^{x^2 + \frac{53}{210}}$.

$\square$

Each of the approximations gets further and further from $n!$ as n tends to infinity. So the quality of the approximations is best judged by considering the percentage error, that is,

$$100 \times \dfrac{|\text{approximation} - n!|}{n!}.$$

In the table S = Stirling, B = Burnside, G = Gosper, M = Mortici, R = Ramanujan, and HV = Hirschhorn and Villarino

n	$n!$	S %error	B %error	G %error	M %error	R % error	HV %error
2	2	4.0	1.7	1.3×10^{-1}	1×10^{-2}	3.3×10^{-3}	1.6×10^{-4}
5	1.2×10^2	1.7	7.6×10^{-1}	2.5×10^{-2}	5.7×10^{-4}	1.2×10^{-4}	1.5×10^{-6}
10	3.6×10^6	8.3×10^{-1}	4.0×10^{-1}	6.6×10^{-3}	7.0×10^{-5}	8.6×10^{-6}	3.0×10^{-8}
20	2.4×10^{18}	4.2×10^{-1}	2.0×10^{-1}	1.7×10^{-3}	8.7×10^{-6}	5.7×10^{-7}	5.2×10^{-10}
50	3.0×10^{64}	1.7×10^{-1}	8.3×10^{-2}	2.7×10^{-4}	5.6×10^{-7}	1.5×10^{-8}	2.3×10^{-12}
100	9.3×10^{157}	8.3×10^{-1}	4.1×10^{-2}	6.9×10^{-5}	6.9×10^{-8}	9.5×10^{-10}	3.6×10^{-14}
10^3	$4.0 \times 10^{2,567}$	8.3×10^{-3}	4.2×10^{-3}	6.9×10^{-7}	6.9×10^{-11}	9.5×10^{-14}	3.7×10^{-20}
10^4	$2.8 \times 10^{35,659}$	8.3×10^{-4}	4.2×10^{-4}	6.9×10^{-9}	6.9×10^{-14}	9.5×10^{-18}	3.7×10^{-26}
10^6	$8.3 \times 10^{5,565,708}$	8.3×10^{-6}	4.2×10^{-6}	6.9×10^{-13}	6.9×10^{-20}	9.5×10^{-26}	3.7×10^{-38}

We calculated the entries in all but the final three rows in the table above using the software package R. The final three rows were calculated using the Wolfram|Alpha software package.

Corollary 4.3 *For x a positive real number,* $\Gamma(x + \frac{1}{2}) \sim \sqrt{2\pi} x^x e^{-x}$.

Proof. Laplace's extension of Stirling's formula Theorem 4.7,

$$\Gamma\left(x+\frac{1}{2}\right) = \Gamma\left(\left(x-\frac{1}{2}\right)+1\right)$$

$$\sim \left(x-\frac{1}{2}\right)^{x-\frac{1}{2}} e^{-(x-\frac{1}{2})} \sqrt{2\pi\left(x-\frac{1}{2}\right)}$$

$$= \left(x-\frac{1}{2}\right)^{x} e^{-x} \sqrt{e} \sqrt{2\pi}$$

$$= x^{x}\left(1-\frac{1}{2x}\right)^{x} e^{-x} \sqrt{e} \sqrt{2\pi}$$

$$= x^{x} e^{-x} \sqrt{2\pi} \left(\left(1-\frac{1}{2x}\right)^{x} \sqrt{e}\right)$$

$$\sim x^{x} e^{-x} \sqrt{2\pi}, \quad \text{as } \lim_{x\to\infty}\left(\left(1-\frac{1}{2x}\right)^{x} \sqrt{e}\right) = 1.$$

□

4.12 Male Births

You should recall that we mentioned that De Moivre started the topic of approximating $n!$ because he was looking at games of chance where it became necessary to approximate $\binom{2n}{n}$ for large values of n. Let us consider now an example which deals with such an estimate.

Example 4.10 Assume that in a particular country there were exactly one million births in 2020. Also assume that the probability that a baby is born biologically male is 0.5. What is the probability in 2020 that there will be exactly 500,000 babies born which are biologically male? And use Stirling's formula to approximate this number. Clearly the probability is $\binom{1,000,000}{500,000}(\frac{1}{2})^{1,000,000}$. These numbers are too big for a calculator. So let us use Stirling formula which says that $n! \sim (n^n/e^n)\sqrt{2\pi n}$. So

$$\binom{2n}{n} = \frac{(2n)!}{n!.n!} \sim \frac{\frac{(2n)^{2n}}{e^{2n}}\sqrt{2\pi.2n}}{\frac{n^n}{e^n}\sqrt{2\pi n}.\frac{n^n}{e^n}\sqrt{2\pi n}} = \frac{2^{2n}}{\sqrt{\pi}\sqrt{n}}$$

Thus $\binom{2n}{n}.2^{-2n} \sim \frac{1}{\sqrt{\pi.n}}.$

So $\binom{1,000,000}{500,000}.2^{-1,000,000} \sim \frac{1}{\sqrt{\pi.500,000}} = 0.00079788\ldots.$

We observe firstly, that Stirling's formula avoided having to calculate 1,000,000!, and $(500,000!)^2$. Secondly we note that the probability that exactly half of those born were biologically male is very small. Finally I mention that the actual value is indeed $0.00079788\ldots.$

4.13 The Basel Problem: Evaluating $\zeta(2) = \sum_{n=1}^{\infty} \dfrac{1}{n^2}$

Shortly we shall evaluate the probability that two randomly chosen natural numbers are coprime, that is, their least common divisor is 1. To answer this, one must first solve the Basel problem which of independent interest in mathematics.

The *Basel problem* was posed by Pietro Mengoli in 1650. Pietro Mengoli (1626–1686) was an Italian mathematician. In 1650 he studied the *alternating harmonic series* and proved that $1 - \dfrac{1}{2} + \dfrac{1}{3} - \cdots + (-1)^{n+1}\dfrac{1}{n} + \cdots = \ln 2$. This can be verified today by (careful) evaluation of the Taylor Series of $\ln(1+x)$.

In 1650, he also posed the problem of evaluating the convergent series $\sum_{n=1}^{\infty} \dfrac{1}{n^2}$.

Mengoli

Despite attempts by the leading mathematicians of the time, this problem remained unsolved for 80 years. In 1734 Leonhard Euler (1707–1783) announced that he had solved the problem. As Euler was born in Basel in Switzerland, the problem became known as the Basel problem. Euler presented his elegant solution to the St. Petersburg Academy of Sciences in 1735. His solution brought him immediate fame, partly because of his youth. Euler's proof in fact relied on an extension of the fundamental theorem of algebra so that he could express $\sin x$ as a particular infinite product, a fact which had not been rigorously proved at that time. In 1741 Euler was able to give a rigorous proof that $\sum_{n=1}^{\infty} \dfrac{1}{n^2} = \dfrac{\pi^2}{6}$.

Euler generalized this to study the ζ function defined by

$$\zeta(s) = \sum_{n=1}^{\infty} \frac{1}{n^s},$$

for $s \in \mathbb{R}$. In 1859 Bernhard Riemann took up Euler's ideas in his paper [25] and extended the zeta function to complex numbers s, proved its basic properties, and discussed between $\zeta(s)$ and the distribution of the prime numbers. The zeta function is known today as the *Riemann zeta function*.

> In his 1859 paper Riemann conjectured that the Riemann zeta function has its zeros only at the negative even integers and complex numbers with real part 1/2.
>
> This conjecture has remained unproved for 160 years and is regarded as one of the most important unsolved problems in mathematics.

4.13 The Basel Problem: Evaluating $\zeta(2) = \sum_{n=1}^{\infty} \frac{1}{n^2}$

It is one of the million-dollar millennium prize problems.

It is of great interest because it would imply significant results about the distribution of prime numbers.

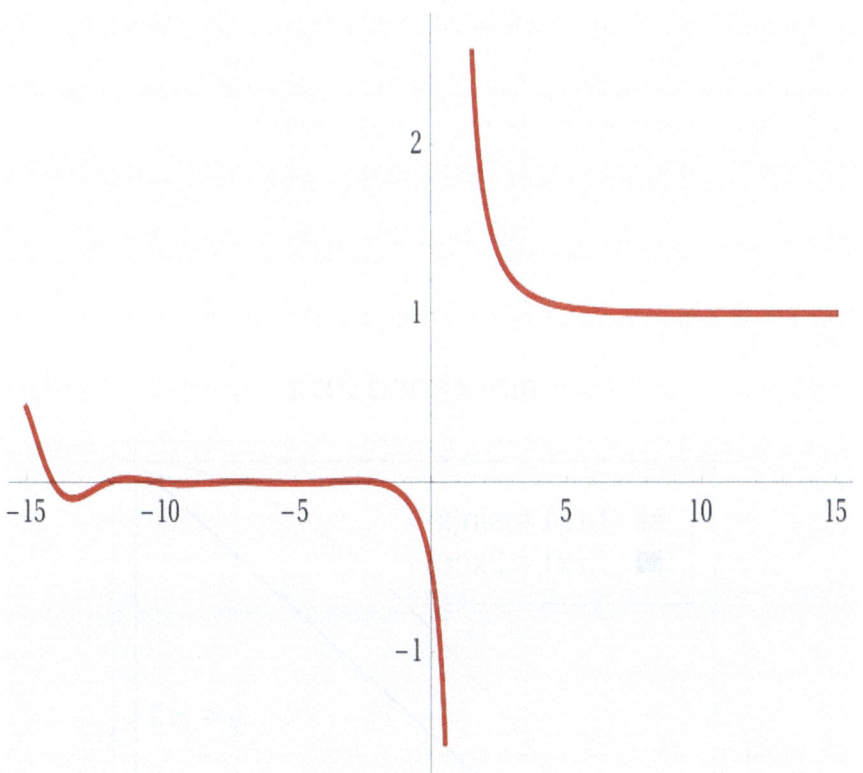

Riemann Zeta Function

Georg Friedrich Bernhard Riemann (1826–1866) was a German mathematician who made contributions to analysis, number theory, and differential geometry. To undergraduates he is known for the rigorous formulation of the integral, the *Riemann integral*. His contributions to complex analysis include the introduction of Riemann surfaces. His famous 1859 paper on the prime-counting function is regarded as one of the most influential papers in analytic number theory.

Riemann

After that digression to put the Basel problem into context, let us proceed to an exposition of the solution. As we said before, there are often many proofs of the same theorem. Some proofs are described as simple proofs, others as informative proofs, and yet others as elementary proofs. An informative proof is one which helps you understand the theorem. A simple proof is one which is usually short, but depends on some other results, sometimes powerful results, which you may or may not know. An elementary proof is one which uses the minimum background knowledge. The shortest proof I have seen on Basel's problem is [15] which claims to be a one sentence proof. Another short proof is in [6]. The exposition here is that of [10] as I regard it as being more informative and is indeed elementary.

Before we begin the proof itself, let us prove a simple fact, namely, that

$$x \in [0, \frac{\pi}{2}] \implies \frac{2}{\pi}x \leq \sin x,$$

which is evident from the graph (but that is not a proof).

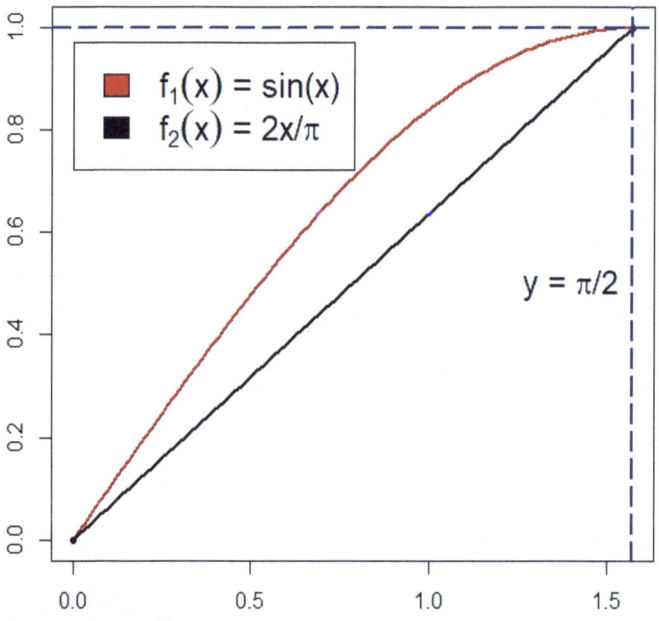

sin(x) and 2x/π

Proof. Note that $\sin(0) = 0$ and $\sin(\frac{\pi}{2}) = 1$. So x and $\frac{\pi}{2}\sin x$ are equal at $x = 0$ and $x = \frac{\pi}{2}$. Put $f(x) = \frac{\sin x}{x}$, for $0 < x \leq \frac{\pi}{2}$.

Then the derivative $f'(x) = \frac{g(x)}{x^2}$, where $g(x) = x \cos x - \sin x$.

4.13 The Basel Problem: Evaluating $\zeta(2) = \sum_{n=1}^{\infty} \frac{1}{n^2}$

So $g'(x) = -x \sin x \leq 0$, for all $x \in (0, \frac{\pi}{2}]$.
Thus g is a decreasing function, for all $x \in (0, \frac{\pi}{2}]$.
Hence $g(x) \leq 0$, for all $x \in [0, \frac{\pi}{2}]$. So $f'(x) \leq 0$ for all $x \in (0, \frac{\pi}{2}]$.
This implies that f is a decreasing function for $x \in (0, \frac{\pi}{2}]$, which in turn says that $\frac{\pi}{2} \cdot \frac{\sin x}{x}$ is a decreasing function, for all $x \in (0, \frac{\pi}{2}]$.
But $\frac{\pi}{2} \cdot \frac{\sin x}{x} = 1$ for $x = \frac{\pi}{2}$.
Hence $\frac{\pi}{2} \cdot \frac{\sin x}{x} \geq 1$, for all $x \in (0, \frac{\pi}{2}]$, which proves the proposition. □

The graph appearing above was created using the R software.
```
f1<-function(x) {
y<- (sin(x))
return(y)        }

value<-c(0,pi/2)
  f2<-c(0,1)
 plot(value,f2, pch=20, ylab="", xlab="")
 fill=c("red","black")
 title(main= expression(paste(sin(x), " and 2x/",pi)),
 col.main="blue",cex.main=1.8)

f3<-function(x) {
y<- (2*x/pi)
return(y)
       }
 curve(f1(x),  xlim=c(0,pi/2), ylim=c(0,2), col="red",
   lwd=2,  add=TRUE)
 curve(f3(x),  xlim=c(0,pi/2), ylim=c(0,2), col="black",
   lwd=2,  add=TRUE)
 legend(0,0.97, legend=c(expression(paste(f[1](x),
 " = sin(x)")),
 expression(paste(f[2](x)," = 2x/",pi))), ,cex=1.6,
 fill=c("red","black"))
 abline(v = pi/2, lty=5, lwd=2, col="blue")
 abline(h = 1, lty=5, lwd=2,col="blue")
 text(1.4,0.5,expression(paste("y = ",pi,"/2")),cex=1.5)
```

Theorem 4.9 $\zeta(2) = \sum_{k=1}^{\infty} \frac{1}{k^2} = \frac{\pi^2}{6}$.

Proof. Define $A_n = \int_0^{\frac{\pi}{2}} (\cos x)^{2n}\, dx$ and $B_n = \int_0^{\frac{\pi}{2}} x^2 (\cos x)^{2n}\, dx$, for $n \in \mathbb{Z}$, $n \geq 0$.

We shall prove that

$$0 \leq \frac{\pi^2}{6} - \sum_{k=1}^{n} \frac{1}{k^2} = 2\frac{B_n}{A_n} \leq \frac{\pi^2}{4(n+1)}. \tag{4.13.1}$$

If we let $n \to \infty$, (4.13.1) implies the statement in the theorem. So we proceed to prove (4.13.1). We use integration by parts and $\sin^2 x = 1 - \cos^2 x$.

$$A_n = \int_0^{\frac{\pi}{2}} \cos x \cos^{2n-1} x\, dx$$

$$= (2n-1) \int_0^{\frac{\pi}{2}} \sin^2 x \cos^{2(n-1)} x\, dx \tag{4.13.2}$$

$$= (2n-1) \int_0^{\frac{\pi}{2}} (1 - \cos^2 x) \cos^{2(n-1)} x\, dx$$

$$= (2n-1)(A_{n-1} - A_n). \tag{4.13.3}$$

Easy manipulation of (4.13.2) and (4.13.3) shows that

$$\int_0^{\frac{\pi}{2}} \sin^2 x \cos^{2(n-1)} x\, dx = \frac{A_n}{2n-1} = \frac{A_{n-1}}{2n}. \tag{4.13..4}$$

Next we express A_n in terms of B_n. We do so by integrating by parts twice. Using it once, we obtain

$$A_n = \int_0^{\frac{\pi}{2}} 1 \times \cos^{2n} x\, dx = 2n \int_0^{\frac{\pi}{2}} x \sin x \cos^{2n-1} x\, dx.$$

Using it a second time, we obtain

$$A_n = -n \int_0^{\frac{\pi}{2}} x^2 \left(\cos x \cos^{2n-1} x - (2n-1) \sin^2 x \cos^{2n-2} x \right) dx$$

$$= -n B_n + n(2n-1) \int_0^{\frac{\pi}{2}} x^2 (1 - \cos^2 x) \cos^{2(n-1)}\, dx$$

$$= -n B_n + n(2n-1)(B_{n-1} - B_n).$$

So $\quad A_n = n(2n-1) B_{n-1} - 2n^2 B_n. \tag{4.13.5}$

Dividing (4.13.5) by $n^2 A_n$ and using (4) we obtain

$$\frac{1}{n^2} = \frac{(2n-1)B_{n-1}}{n A_n} - \frac{2B_n}{A_n} = \frac{2B_{n-1}}{A_{n-1}} - \frac{2B_n}{A_n}.$$

As this is true for all integers $n \geq 1$, we can sum both sides to get for all $n \geq 1$

$$\sum_{k=1}^{n} \frac{1}{k^2} = \sum_{k=1}^{n} \left(\frac{2B_{k-1}}{A_{k-1}} - \frac{2B_k}{A_k} \right)$$

$$= \frac{2B_0}{A_0} - \frac{2B_n}{A_n}, \quad \text{by noting the intermediate terms in the summation cancel.}$$

As $A_0 = \int_0^{\frac{\pi}{2}} 1\, dx = \frac{\pi}{2}$ and $B_0 = \int_0^{\frac{\pi}{2}} x^2 dx = \frac{\pi^3}{24}$, we see that $\frac{2B_0}{A_0} = \frac{\pi^2}{6}$.

Thus we have for all $n \geq 1$, $\sum_{k=1}^{n} \frac{1}{k^2} = \frac{\pi^2}{6} - \frac{2B_n}{A_n}$. \hfill (4.13.6)

Now we have $B_n = \int_0^{\frac{\pi}{2}} x^2 (\cos x)^{2n}\, dx$

$$\leq \left(\frac{\pi}{2}\right)^2 \int_0^{\frac{\pi}{2}} \sin^2 x \cos^{2n} x\, dx$$

$$= \frac{\pi^2}{4} \cdot \frac{A_n}{2(n+1)} \quad \text{by replacing } n \text{ by } n+1 \text{ in (4.13.4).} \quad (4.13.7)$$

(4.13.6) and (4.13.7) then yield the required result (4.13.1) □

4.14 Probability that Two Randomly Chosen Natural Numbers Are Coprime

While the idea of this problem is very easily understood, the verification that the problem is meaningful is not trivial as one must explain what one means by two randomly chosen natural numbers and show that this is indeed possible. An equivalent formulation of the problem is as follows: what is the probability that a fraction with randomly selected numerator and denominator from the set of natural numbers is irreducible, that is, the numerator and denominator have no common factor > 1. Who first stated this problem is not definitely known, but it probably dates from about 200–250 years ago. For a history of the problem, from a Russian perspective at least, see [1]. In the literature there are a number of solutions to this problem, none of which could be described as easy. A standard proof can be found as [12, Theorem 332]. The proof presented here is elementary. It is based on Theorem 4.9 and [28].

Definition 4.10 If $a, b, c, n, p \in \mathbb{Z}$, then c is said to be a *divisor* of n if there exists an $m \in \mathbb{N}$ such that $n = mc$. The number p is said to be a *prime number* if it has precisely 2 divisors which are positive integers. A positive integer which is not a prime number and is strictly greater than 1 is said to be a *composite number*. The number c is said to be the *greatest common divisor* of a and b if c is a divisor of a and b and no number greater than c is a divisor of both a and b. The greatest common divisor of a and b is written (a, b). The numbers a and b are said to be *coprime* (or *relatively prime* or *mutually prime*) if $(a, b) = 1$.

Example 4.11 The numbers $2, 3, 5, 7, 11, 13, 17, 19 \ldots$ are prime numbers, while $2, 4, 6, 8, 9, 10, 12, \ldots$ are not prime numbers. 1 is not a prime number as it has only 1 divisor which is a positive integer. Clearly 7 and 13 are coprime, as are 14 and 39. We see that if either (or both) of a and b are prime numbers, then a and b are coprime. On the other hand, a and b coprime does not imply either a or b are prime numbers.

Existence of an Infinite Number of Prime Numbers

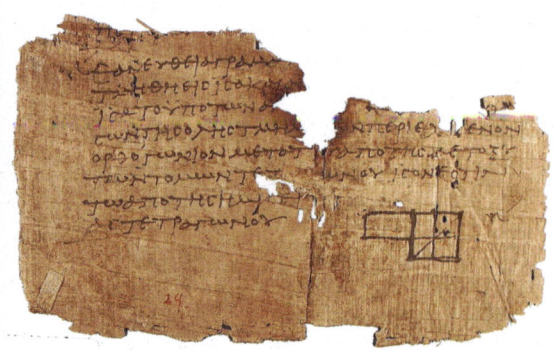

Joseph Durham's Euclid   Fragment of Papyrus of Euclid's Elements

It is known that there is an infinite number of prime numbers. The first known proof of this fact is attributed to Euclid who was an influential Greek mathematician who lived in Alexandria about 2,300 years ago. Several alternative proofs are presented in [4].

Euclid is regarded as the father of geometry. His work *Elements* developed geometry from a set of axioms. (His geometry is now called Euclidean geometry.) This book was used to teach mathematics until about 130 years ago. Pictured above is a photo of a statue of Euclid by the English sculptor Joseph Durham (1814–1877) and is located in the Oxford University Museum of Natural History. The fragment of papyrus also pictured above has a diagram which accompanies Proposition 5 of Book II of the Elements. The fragment is dated at about 1900 years ago. It was discovered by the English Egyptologist Bernard Pyne Grenfell (1869–1926) and the English Papyrologist Arthur Surridge Hunt (1871–1934) at Oxyrhynchus (later known as Al-Bahnasa), a city in Middle Egypt located about 160 km south-southwest of Cairo.

We next present Euclid's proof that the number of prime numbers is infinite.

Proof by Contradiction Begins with Suppose

The proof is done by a method known as *proof by contradiction*. We begin by *supposing* that what we would like to prove is *false*. We continue until we reach a conclusion which we know is untrue. So we have a *contradiction*. Therefore what we supposed must be false.

So that you know I am using this proof by contradiction method of proof, I will always begin with *suppose*. I shall not use the word suppose in any other context. Instead I will probably use the word assume.

> From the definition of prime number, every positive integer >1 is either a prime number or is divisible by a prime number.

Theorem 4.10 *There is an infinite number of prime numbers.*

Proof. Suppose that there is only a finite number of prime numbers $p_1, p_2, \ldots, p_n$, where $n \in \mathbb{N}$.

Consider the positive integer $N = p_1.p_2.\ldots.p_n + 1$. This number is bigger than each of $p_1, p_2, \ldots, p_n$. So it cannot be a prime number, as by supposition these are the only prime numbers. But N must be divisible by a prime number.

But if you divide the number $N = p_1.p_2.\ldots.p_n + 1$ by any of $p_1, p_2, \ldots, p_n$, there is a remainder of 1. So N is not a prime number and is not divisible by a prime number and it is strictly bigger than 1. This is a *contradiction*. So our supposition was false, and there must exist an infinite number of prime numbers. □

Infinite Products

Definition 4.11 If $x_1, x_2, \ldots, x_n \in \mathbb{R}$, for $n \in \mathbb{N}$, then $\prod_{i=1}^{n} x_i$ is defined to be equal to the product $x_1.x_2.\ldots.x_n$. Further, the *infinite product* $\prod_{i=1}^{\infty} x_i$ is defined to be $\lim_{n \to \infty} \prod_{i=1}^{n} x_i$, if that limit exists.

Proposition 4.4 *Let $p_1, p_2, \ldots p_n, \ldots$ be the prime numbers in ascending order. Then $\prod_{n=1}^{\infty}\left(1 - \frac{1}{p_n^2}\right)$ exists; that is, $\lim_{n\to\infty} \prod_{i=1}^{n}\left(1 - \frac{1}{p_i^2}\right)$ exists.*

Proof. Consider $a_n = \prod_{i=1}^{n}\left(1 - \frac{1}{p_i^2}\right)$. Observe that a_n is the product of n terms $\left(1 - \frac{1}{p_i^2}\right)$, each of which is positive and strictly less than 1. This implies that $\lim_{n\to\infty} a_n$ exists. $\square$

Shortly we will actually evaluate this product.

Theorem 4.11 $\zeta(2) = \sum_{n=1}^{\infty} \frac{1}{n^2} = \prod_{k=1}^{\infty} \frac{1}{1 - \frac{1}{p_k^2}}$, *where $p_1, p_2, \ldots, p_k, \ldots$ are the prime numbers in ascending order.*

Proof.

$$\prod_{k=1}^{\infty} \frac{1}{1 - \frac{1}{p_k^2}} = \left(\frac{1}{1 - \frac{1}{p_1}^2}\right)\left(\frac{1}{1 - \frac{1}{p_2}^2}\right)\cdots\left(\frac{1}{1 - \frac{1}{p_k}^2}\right)\cdots$$

$$= \left(1 + \frac{1}{p_1^2} + \frac{1}{p_1^4} + \cdots + \frac{1}{p_1^{2k}} + \cdots\right)$$

$$\times \left(1 + \frac{1}{p_2^2} + \frac{1}{p_2^4} + \cdots + \frac{1}{p_2^{2k}} + \cdots\right)\cdots$$

$$= 1 + \sum_{i\geq 1} \frac{1}{p_i^2} + \sum_{j\geq i\geq 1} \frac{1}{p_i^2 p_j^2} + \sum_{k\geq j\geq i\geq 1} \frac{1}{p_i^2 p_j^2 p_k^2} + \cdots$$

$$= 1 + \frac{1}{2^2} + \frac{1}{3^2} + \cdots \frac{1}{n^2} + \cdots$$

$$= \sum_{n=1}^{\infty} \frac{1}{n^2} = \zeta(2).$$

$\square$

The Corollary below follows immediately from Theorem 4.11 and the Basel Theorem 4.9.

Corollary 4.4 $\frac{1}{\zeta(2)} = \frac{6}{\pi} = \prod_{n=1}^{\infty}\left(1 - \frac{1}{p_n^2}\right).$

4.14 Probability that Two Randomly Chosen Natural Numbers Are Coprime

Definition 4.12 If a is any positive real number, then $[a]$ is said to be the *integer part* of a and is defined to be the largest integer such that $[a] \leq a$.

Example 4.12 $[n] = n$ for any positive integer while $[3.6] = 3$.

Lemma 4.1 *Let N be any positive integer. Then*

$$1 + \frac{1}{2} + \frac{1}{3} + \cdots + \frac{1}{N} < 2\sqrt{N}.$$

Proof.

$$1 + \frac{1}{2} + \frac{1}{3} + \cdots + \frac{1}{N}$$

$$= \left(1 + \frac{1}{2} + \frac{1}{3} + \cdots + \frac{1}{[\sqrt{N}]}\right) + \left(\frac{1}{[\sqrt{N}]+1} + \frac{1}{[\sqrt{N}]+2} + \cdots + \frac{1}{N}\right).$$

The first parenthesis consists of $[\sqrt{N}]$ terms each of which is ≤ 1. Therefore its value is $\leq [\sqrt{N}] \leq \sqrt{N}$.

The second parenthesis has $N - [\sqrt{N}]$ terms all $< \frac{1}{\sqrt{N}}$ terms. Therefore is value is $< \frac{(N-[\sqrt{N}])}{\sqrt{N}} < \frac{N}{\sqrt{N}} = \sqrt{N}$.

So the sum of the two parentheses is $< 2\sqrt{N}$, which proves the Lemma. □

Lemma 4.2 *Let N be any positive integer and m any positive integer strictly greater than 1. Then*

$$\frac{1}{(N+1)^2} + \frac{1}{(N+2)^2} + \frac{1}{(N+3)^2} + \cdots + \frac{1}{(N^m)^2} < \frac{1}{N}.$$

Proof.

$$\frac{1}{(N+1)^2} + \frac{1}{(N+2)^2} + \frac{1}{(N+3)^2} + \cdots + \frac{1}{(N^m)^2}$$

$$< \frac{1}{N(N+1)} + \frac{1}{(N+1)(N+2)} + \frac{1}{(N+2)(N+3)} + \cdots + \frac{1}{(N^m-1)N^m}$$

$$= \left(\frac{1}{N} - \frac{1}{N+1}\right) + \left(\frac{1}{N+1} - \frac{1}{N+2}\right) + \cdots + \left(\frac{1}{N^m-1} - \frac{1}{N^m}\right)$$

$$= \frac{1}{N} - \frac{1}{N^m} < \frac{1}{N}.$$

□

Lemma 4.3 *Let N be any positive integer and m a positive integer strictly greater than 1. Then*

$$\left(\left(\frac{N}{2}\right)^2 - \left[\frac{N}{2}\right]^2\right) + \left(\left(\frac{N}{3}\right)^2 - \left[\frac{N}{3}\right]^2\right) + \cdots + \left(\left(\frac{N}{N^m}\right)^2 - \left[\frac{N}{N^m}\right]^2\right)$$
$$< N^2 \left(\frac{2}{N}\left(1 + \frac{1}{2} + \cdots + \frac{1}{N}\right) + \frac{1}{(N+1)^2} + \frac{1}{(N+2)^2} + \cdots + \frac{1}{(N^m)^2}\right) \quad (4.14.1)$$

Proof. Let $r \in \mathbb{N}$. If $r > N$, then $\frac{N}{r} < 1$ and so

$$r > N \implies \left[\frac{N}{r}\right] = 0. \quad (4.14.2)$$

If $r \leq N$, we note that

$$\left[\frac{N}{r}\right] > \frac{N}{r} - 1$$

and that both sides of this inequality are ≥ 0. So we can square both sides of the inequality to obtain

$$\left[\frac{N}{r}\right]^2 > \left(\frac{N}{r}\right)^2 - 2\left(\frac{N}{r}\right) + 1$$

and so

$$r \leq N \implies \left(\frac{N}{r}\right)^2 - \left[\frac{N}{r}\right]^2 < 2\left(\frac{N}{r}\right) - 1 < 2\left(\frac{N}{r}\right). \quad (4.14.3)$$

Applying (4.14.2) and (4.14.3) to the left hand side of (4.14.1) yields

$$\left(\left(\frac{N}{2}\right)^2 - \left[\frac{N}{2}\right]^2\right) + \left(\left(\frac{N}{3}\right)^2 - \left[\frac{N}{3}\right]^2\right) + \cdots + \left(\left(\frac{N}{N^m}\right)^2 - \left[\frac{N}{N^m}\right]^2\right)$$
$$< \frac{2N}{1} + \frac{2N}{2} + \cdots + \frac{2N}{N} + \left(\frac{N}{N+1}\right)^2 + \left(\frac{N}{N+2}\right)^2 + \cdots + \left(\frac{N}{N^m}\right)^2$$
$$= N^2 \left(\frac{2}{N}\left(1 + \frac{1}{2} + \cdots + \frac{1}{N}\right) + \frac{1}{(N+1)^2} + \frac{1}{(N+2)^2} + \cdots + \frac{1}{(N^m)^2}\right)$$

□

Lemma 4.4 *Let N be any positive integer and m a positive integer strictly greater than 1. Put*

$$a_m = \left(\left(\frac{N}{2}\right)^2 - \left[\frac{N}{2}\right]^2\right) + \left(\left(\frac{N}{3}\right)^2 - \left[\frac{N}{3}\right]^2\right) + \cdots + \left(\left(\frac{N}{N^m}\right)^2 - \left[\frac{N}{N^m}\right]^2\right)$$

and $b_m = \frac{1}{N^2} a_m$. Then $\lim_{N \to \infty} b_m = 0$.

4.14 Probability that Two Randomly Chosen Natural Numbers Are Coprime

Proof. By Lemmas 4.3, 4.2, and 4.1 $b_m < \dfrac{4}{\sqrt{N}} + \dfrac{1}{N}$, from which the Lemma follows.

□

Lemma 4.5 *Let N be any positive integer and consider all the prime numbers $p_1, p_2, \ldots, p_m \leq N$, so that $2 = p_1 < p_2 < \cdots < p_m$. Put*

$$c_N = \left(\frac{N}{p_1}\right)^2 - \left[\frac{N}{p_1}\right]^2 + \cdots + \left(\frac{N}{p_m}\right)^2 - \left[\frac{N}{p_m}\right]^2$$
$$+ \left(\frac{N}{p_1 p_2}\right)^2 - \left[\frac{N}{p_1 p_2}\right]^2 + \cdots + \left(\frac{N}{p_1 p_2 \cdots p_m}\right)^2 - \left[\frac{N}{p_1 p_2 \cdots p_m}\right]^2$$

and $d_N = \frac{1}{N^2} c_N$. Then $\lim\limits_{N \to \infty} d_N = 0$.

Proof. Observing that every term in c_N also appears as a term in a_m in Lemma 4.4, it follows that $d_N \leq b_m$. Further, as $d_N \geq 0$, Lemma 4.4 implies that $\lim\limits_{N \to \infty} d_N = 0$.

□

As a corollary to Lemma 4.5, we have as follows.

Lemma 4.6 *Let N be any positive integer and consider all the prime numbers $p_1, p_2, \ldots, p_m \leq N$, so that $2 = p_1 < p_2 < \cdots < p_m$. Put*

$$e_N = \left(\frac{N}{p_1}\right)^2 + \cdots + \left(\frac{N}{p_m}\right)^2 + \left(\frac{N}{p_1 p_2}\right)^2 + \cdots + \left(\frac{N}{p_1 p_2 \cdots p_m}\right)^2$$

and

$$f_N = \left[\frac{N}{p_1}\right]^2 + \cdots + \left[\frac{N}{p_m}\right]^2 + \left[\frac{N}{p_1 p_2}\right]^2 + \cdots + \left[\frac{N}{p_1 p_2 \cdots p_m}\right]^2.$$

If $\lim\limits_{N \to \infty} \dfrac{e_N}{N^2}$ exists and equals $s \in \mathbb{R}$, then $\lim\limits_{N \to \infty} \dfrac{f_N}{N^2}$ exists and equals s.

Lemma 4.7 *If f_N is defined as in Lemma 4.6, $\lim\limits_{N \to \infty} \dfrac{f_N}{N^2}$ exists.*

Proof. It is readily seen that $\dfrac{e_N}{N^2} = 1 - \left(1 - \dfrac{1}{p_1^2}\right)\left(1 - \dfrac{1}{p_2^2}\right) \cdots \left(1 - \dfrac{1}{p_m^2}\right).$

As N increases, the number of factors in the product increases and each factor is < 1. So as N increases, $\dfrac{e_N}{N^2}$ decreases but remains ≥ 0. This implies that $\lim\limits_{N \to \infty} \dfrac{e_N}{N^2}$ exists. Then by Lemma 4.3, $\lim\limits_{N \to \infty} \dfrac{f_N}{N^2}$ exists.

□

As observed previously, we have as follows.

Lemma 4.8 *With the same notation as in Lemma 4.6,*

$$\frac{e_N}{N^2} = 1 - \left(1 - \frac{1}{p_1^2}\right)\left(1 - \frac{1}{p_2^2}\right) \cdots \left(1 - \frac{1}{p_m^2}\right).$$

4.15 Principle of Inclusion and Exclusion

Remark 4.5 We saw previously that for finite sets A, B, C,

$$|A \cup B \cup C| = |A| + |B| + |C| - |A \cap B| - |A \cap C| - |B \cap C| + |A \cap B \cap C|$$

where $|S|$ denotes the number of elements in any set S. This geralizes to n finite sets in what is known as the *Principle of Inclusion and Exclusion* as follows:
If $A_1, A_2, \ldots, A_n$ are finite sets for any $n \in \mathbb{N}$, then

$$\left| \bigcup_{i=1}^{n} A_i \right| = \sum_{i=1}^{n} |A_i| - \sum_{1 \le i < j \le n} |A_i \cap A_j| + \sum_{1 \le i < j < k \le n} |A_i \cap A_j \cap A_k| - \ldots + (-1)^{n-1} |A_1 \cap A_2 \cap \cdots \cap A_n|.$$

Finally we address directly the problem of the probability that two natural numbers chosen at random have what probability of being coprime.

As we said earlier, we have to say what we mean by saying that two natural numbers chosen at random have a certain probability of being coprime. The next proposition is key to clarifying this matter.

Proposition 4.5 *Let N be a positive integer and let a, b be chosen at random (with replacement) from the finite set $\{1, 2, \ldots, N\}$. Let s_N be the probability that these randomly chosen integers are coprime. Then $\lim_{N \to \infty} s_N = s$ exists and equals*

$$\lim_{N \to \infty} \left(1 - \frac{f_N}{N^2}\right) = \lim_{N \to \infty} \left(1 - \frac{e_N}{N^2}\right) = \prod_{n=1}^{\infty} \left(1 - \frac{1}{p_n^2}\right),$$

where f_N and e_N is as in Lemma 4.6 and $p_1, p_2, \ldots, p_n, \ldots$ is the set of all prime numbers in ascending order.

Proof. Firstly let us make clear what we mean by *chosen at random* here means. It means that the probability of choosing any of the integers between 1 and N is $\frac{1}{N}$. As is obvious, this could not work as a definition if instead of the finite set $\{1, 2, \ldots, N\}$ we chose the infinite set $\mathbb{N}$. We wish to determine the probability that the randomly chosen $a, b \in \{1, 2, \ldots, N\}$ are coprime.

a and b are coprime $\iff$ no prime number p divides both a and b.

So a and b are not coprime $\iff$ p_1 or p_2 or $\ldots p_m$ divide both a and b, where $p_1, p_2, \ldots, p_m$ is the set of all primes satisfying $p_1 < p_2 < \ldots p_m \le N$.

Let A_i be the set of ordered pairs of integers between 1 and N which are divisible by p_i. Then $A_1 \cup A_2 \cup \cdots \cup A_n$ is the set of all ordered pairs (a, b) which are divisible by one or more of the p_i; that is it is the set of ordered pairs (a, b) such that a and b are not coprime.

We shall be applying the principle of inclusion and exclusion. First, $|A_i|$ is the number of ordered pairs (a, b) which are divisible by p_i. The number of multiples

4.15 Principle of Inclusion and Exclusion

of p_i between 1 and N is $\left[\dfrac{N}{p_i}\right]$. So $|A_i| = \left[\dfrac{N}{p_i}\right]^2$. Next, we see that $A_i \cap A_j$ is the set of ordered pairs (a,b) such that a and b are multiples of p_i and p_j; that is, they are multiples of $p_i p_j$. Since there are $\left[\dfrac{N}{p_i p_j}\right]$ such multiples, $|A_i \cap A_j| = \left[\dfrac{N}{p_i p_j}\right]^2$. Similarly $|A_i \cap A_j \cap A_k| = \left[\dfrac{N}{p_i p_j p_k}\right]^2$ and so on. Thus by the principle of inclusion and exclusion, the number of ordered pairs (a,b) where a and b are not coprime is equal to f_N of Lemma 4.6. The total number of ordered pairs (a,b) such that a and b are between 1 and N is N^2. So the probability s_N that an ordered pair (a,b) is coprime is $\dfrac{N^2 - f_N}{N^2} = 1 - \dfrac{f_N}{N^2}$. By Lemma 4.7 $\lim_{N\to\infty} \dfrac{f_N}{N^2}$ exists, and so $\lim_{N\to\infty} s_N = s$ exists. The proposition follows then from Lemma 4.8. □

Definition 4.13 The probability that two natural numbers a and b chosen randomly are coprime is defined to be s in Proposition 4.5

Theorem 4.12 *The probability that two natural numbers a and b chosen randomly are coprime equals* $\dfrac{1}{\zeta(2)} = \dfrac{6}{\pi^2} = 0.6079\ldots$.

Proof. The theorem follows immediately from Corollary 4.4, Proposition 4.5, and Definition 4.13. □

Remark 4.6 There are proofs of Theorem 4.12 in the literature which seem to be much shorter and simpler than the one presented here or the one in [12]. For example [3] is a lovely short note which purports to prove this result. Now the paper is lovely in that it does suggest nicely what the probability should be. However, like various other proofs in the literature, it is not rigorous. For a rigorous version of the approach in [3], see [23].

Remark 4.7 [12, Theorem 333] proves an allied result to Theorem 4.12, namely, that the probability that a natural number is quadratfrei is also $\dfrac{6}{\pi^2}$, where a natural number is said to be *quadratfrei* if it is not divisible by the square of any prime number.

It is natural to ask what is the probability p_n that $n > 2$ randomly chosen natural numbers are coprime. This is shown to be $\dfrac{1}{\zeta(n)}$; see [21, 22]. We note that $p_3 = \dfrac{1}{\zeta(3)} = 0.8319\ldots$, $p_4 = 0.9239\ldots$, $p_{10} = 0.9990\ldots$ and $\lim_{n\to\infty} p_n = \lim_{n\to\infty} \dfrac{1}{\zeta(n)} = 1$. More general results appeared in the PhD thesis of my Australian colleague Bob Buttsworth and in his publication [8].

$$\dfrac{1}{\zeta(x)}$$

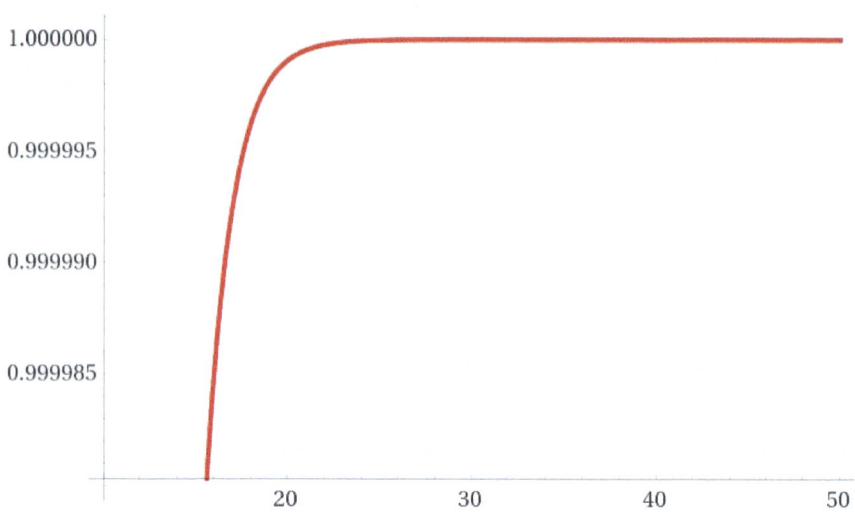

Carefree Couples

It may seem curious that the probability that two randomly chosen natural numbers are coprime is $\dfrac{6}{\pi^2}$ which is equal to the probability that a natural number is quadratfrei. We might ask what is the probability that two randomly chosen natural numbers are both quadratfrei and coprime. Our first guess might be that the probability is $\dfrac{1}{\zeta(2)} \cdot \dfrac{1}{\zeta(2)} \cdot \dfrac{1}{\zeta(2)} = 0.2246\ldots$. However, in [14] such a and b are said to be *strongly carefree couples*, and Moree proves that the probability a and b are strongly carefree is $0.2867\ldots$.

Problems

4.1 Using the software package R, evaluate Stirling's approximation of 40! and evaluate the % error.

4.2 Consider the following sequences and say which are divergent, which have limit ∞ or limit $-\infty$, which are convergent, and if convergent what is their limit.

  (i) $1, 0, 1, 0, 1, 0, \ldots 1, 0, \ldots$
  (ii) $2^1, 2^2, 2^3, \ldots 2^n, \ldots$

4.15 Principle of Inclusion and Exclusion

(iii) $a^1, a^2, a^3, \ldots, a^n, \ldots$, where $a < 1$.
(iv) $1 + (-1)^n$, where $n = 1, 2, 3, \ldots$
(v) $\dfrac{1 + (-1)^n}{n}$, where $n = 1, 2, 3, \ldots$.

4.3 Using L'Hopital's rule, evaluate each of the following limits:

(i) $\lim\limits_{x \to \infty} \dfrac{\ln x}{x}$. (Recall $\dfrac{d(\ln x)}{dx} = \dfrac{1}{x}$.)

(ii) $\lim\limits_{x \to \infty} \dfrac{x \ln x}{e^x}$. (Use L'Hopital's rule twice.)

(iii) Using mathematical induction, prove that $\lim\limits_{x \to \infty} e^{-x} x^n = 0$, for $n \in \mathbb{N}$.

(iv) Deduce from (iii) that for any real number a, $\lim\limits_{x \to \infty} e^{-x} x^a = 0$.

4.4 Let h_1, h_2, h_3 be functions from (a, ∞) to $\mathbb{R} \setminus \{0\}$, where $a \in \mathbb{R}$. If $\lim\limits_{x \to \infty} h_1(x) = 1$, $\lim\limits_{x \to \infty} h_2(x) = 1$, and $\lim\limits_{x \to \infty} h_3(x) = \infty$, show that $h_3(x)h_1(x) \sim h_3(x)h_2(x)$.

4.5 Using Problem 4.4 show that if $f(x) = 3x^2 + 2x + 100$ and $g(x) = 3x^2 - 1{,}000x - 100$, then $f(x) \sim g(x)$.

4.6 Verify that if $f(x) = e^x + x^2 - 100$ and $g(x) = e^x - 1{,}000x^5$, then $f(x) \sim g(x)$. (You may assume that if $f(x)$ is any polynomial, then $f(x) \neq 0$ for $x > a$, for some $a \in \mathbb{R}$, and $\lim\limits_{x \to \infty} \dfrac{f(x)}{e^x} = 0$.)

4.7 Verify that if f and g are two functions from $\mathbb{R}$ to $\mathbb{R}$ such that $g(x) = f(x)(1 - x^3)$ and $\lim\limits_{x \to \infty} f(x) = \infty$, then $f(x) \sim g(x)$.

4.8 Using the Comparison Test for Infinite Integrals Theorem 4.2 and Problem 4.3(iv), prove that $\int_1^\infty e^{-x} x^a \, dx$ converges for each real number a.
[Hint: Note $\int_1^\infty x^{-2} \, dx$ converges.]

4.9 Verify the statements in Corollary 4.1 (ii)–(iv).

4.10 Previously we saw that using the software package R, we could prove that the area under the standard normal distribution curve between $x = -3$ and $x = 3$ is $0.9973\ldots$. What is the area under the curve (i) between $x = -2$ and $x = 2$ and (ii) between $x = -4$ and $x = 4$?

4.11 Prove the statements in Proposition 4.2

4.12 Prove the statement in Example 4.1(xi). [Hint. Consider the greatest lower bound of the sequence.]

Ernest William Barnes (1874–1953) was an English mathematician and scientist who left mathematics and in 1924 became Bishop of Birmingham. While at Cambridge University, he was assigned the task of being the tutor of Srinivasan Ramanujan. We have seen that Ramanujan proved a much better asymptotic formula for $n!$, Interestingly Barnes discovered an asymptotic formula for $\Gamma(x + \frac{1}{2})$ which is not well-known. For suitable $c_n, d_n \in \mathbb{R}$,

Barnes

$$\Gamma\left(x + \frac{1}{2}\right) \sim \sqrt{2\pi} x^x e^{-x} \exp\left(-\frac{1}{24x} + \frac{7}{2,880x^3} - \frac{31}{40,320x^5} + \cdots + \frac{c_n}{d_n x^n}\right).$$

4.13 Using Corollary 4.3, verify the above asymptotic formula for

$$\Gamma\left(x + \frac{1}{4}\right) \sim \sqrt{2\pi} x^x e^{-x} \left(x - \frac{3}{4}\right)^{-\frac{1}{4}}.$$

De Montmort's Matching Problem

As the last problem in this book, we shall discuss the de Montmort matching problem which was proposed in 1708 by the French mathematician Pierre Remond de Montmort (1678–1719).

He was elected a Fellow of the Royal Society in 1715 and became a member of the French Academy of Sciences in 1716.

He introduced the notion of a *derangement*, which is a permutation of the elements of a set, such that no element appears in its original position; that is a permutation with no fixed point.

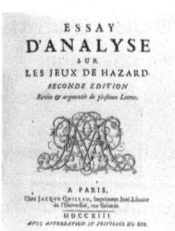
de Montmort

He is also known for his book on probability and games of chance, "Essay d'analyse sur les jeux de hazard" and for naming Pascal's triangle after Blaise Pascal (1623–1662), a French mathematician, physicist, inventor, and philosopher.

Blaise Pascal

4.14 An absent-minded professor (no names mentioned) wrote n letters and sealed then in n envelopes without writing the addresses on the envelopes. Having forgotten which letter he put in which envelope, he wrote the n addresses on the envelopes randomly.

(i) What is the probability p_n that at least one of the letters is addressed correctly? (It is surprising to note that for $n > 2$, p_n is between 0.6 and $\frac{2}{3}$.)

(ii) Prove that $\lim_{n \to \infty} p_n = 1 - \frac{1}{e}$.

[Hint: Show firstly the probability that the ith envelope has the correct letter in it is $\frac{1}{n}$. Next show that the probability that the ith and the jth envelopes have the right letter in them is $\frac{1}{n(n-1)}$.]

4.16 Credit for Images

- Stirling's Methodus differentialis, 1764. Public Domain.
- Plaque to the Stirlings of Garden, Dunblane Cathedral. Licensed under the Creative Commons Attribution-Share Alike 4.0 International license. https://commons.wikimedia.org/wiki/File:Plaque_to_the_Stirlings_of_Garden, Dunblane_Cathedral.jpg
- Portrait of Abraham de Moivre. Public Domain.
- The Doctrine of Chances. Public Domain.
- L'Hôpital's textbook. Public Domain.
- Guillaume de l'Hôpital. Public Domain.
- Johann Bernoulli. Public Domain.
- Nicole Oresme. Public Domain.
- Gottfried Wilhelm (von) Leibniz. Public Domain.
- Johann Carl Friedrich Gauss. Public Domain.
- Leonhard Euler. Public Domain.
- Siméon Denis Poisson. Public Domain.
- Daniel Bernoulli. Public Domain.
- Agner Krarup Erlang. Public Domain.
- Friedrich Robert Helmert. Public Domain.
- Karl Pearson. Public Domain. Public Domain.
- Brook Taylor. Public Domain.
- Colin Maclaurin. Public Domain.
- Sergei Bernstein. Public Domain.
- Paul Erd"os playing Go, November 1979. Copyright held by Sidney A. Morris.
- Euclid statue by Joseph Durham. Public Domain.
- Fragment of Papyrus from Euclid's Elements. Public Domain.
- Georg Friedrich Bernhard Riemann. Public Domain.
- Pietro Mengoli. Public Domain
- Michael Hirschhorn. Permission from Hirschhorn to use the photo.
- William Burnside. Public Domain.

- Ralph William Gosper Jr. Creative Commons Attribution 2.0 Generic license. https://commons.wikimedia.org/wiki/File:Bill_Gosper_2006.jpg
- Burnide's Theory of Probability. Accessed from Internet Archive. Public Domain.
- Huygens book. Public Domain.
- Cristiaan Huygens. Public Domain.
- Srinivasa Ramanujan. Copyright free. https://commons.wikimedia.org/wiki/FileSrinivasa_Ramanujan_-_OPC_-_1.jpg
- Ernest William Barnes. Public Domain.
- de Montmort's book. Public Domain.
- Blaise Pascal. Permission is granted to copy, distribute and/or modify this document under the terms of the GNU Free Documentation License. https://commons.wikimedia.org/wiki/File:Blaise_Pascal_Versailles.JPG

References

1. Abramovich, A., Nikitin, Y.Y.: On the probability of co-primality of numbers chosen at random: From Euler identity to Haar measure on the ring of adeles. Bernoulli News **24**, 7–13 (2017)
2. Abramowitz, M., Stegun, I.: Handbook of Mathematical Functions with Formaulas, Graphs, and Mathematical Tablea, US Department of Commerce, National Bureau of Standards (1964)
3. Abrams, A.D., Paris, M.J.: The probability that (a,b)=1. College Math. J. **23**, 47 (1992)
4. Aigner, M., Ziegler, G.M.: Proofs from the Book, 6th edn. with Illustration by K.H. Hofmann. Springer, Berlin (2018)
5. Apostol, T.M.: Calculus, vol. 1, 2nd edn. Wiley, New York (1967)
6. Apostol, T.M.: A proof that euler missed: evaluating $\zeta(2)$ the easy way. Math. Intell. **5**, 59–60 (1983). https://doi.org/10.1007/BF03026576
7. Burnside, W.: A rapidly converging series for N!. Messenger Math. **46**, 157–159 (1917)
8. Buttsworth, R.N.: On the probability that given polynomials have a specified highest common factor. J. Number Theory **12**, 487–498 (1983). Corrigenda J. Number Theory **16**, 283, (1980)
9. Chen, C.-P.: A more accurate approximation for the gamma function. J. Number Theory **164**, 417–428 (2016)
10. Daners, D.: A rapidly converging series for log N!. Math. Mag. **85**, 361–364 (2012)
11. Gosper, R.W.: Decision procedure for indefinite hypogeometric summation. Proc. Nat. Acad. Sci. USA **75**, 40–42 (1978)
12. Hardy, G.H., Wright, E.M.: An Introduction to the Theory of Numbers, 5th edn. Clarendon Press, Oxford (1979)

13. Hirschhorn, M., Villarino, M.B.: A refinement of Ramanujan's factorial approximation. Ramanujan J. **34**, 73–81 (2014). https://doi.org/10.1007/s11139-013-9494-y
14. Moree, P.: Counting carefree couples (2014). https://arxiv.org/pdf/math/0510003
15. Moreno, S.G.: A One-Sentence and Truly Elementary Proof of the Basel Problem (2015). https://arxiv.org/abs/1502.07667
16. Morris, S.A.: Tweaking Ramanujan's approximation of n!. Fund. J. Math. Appl. **5**(1), 10–15 (2022). https://dx.doi.org/10.33401/fujma.995150
17. Mortici, C.: A substantial improvement of the Stirling formula. Appl. Math. Lett. **24**, 1351–1354 (2011)
18. Mortici, C.: On Gosper's formula for the gamma function. J. Math. Inequal. **25**, 611–614 (2011)
19. Namias, V.: A simple derivation of Stirling's asymptotic series. Am. Math. Monthly **95**, 161–169 (1986)
20. Nemes, G.: On the coefficients of the asymptotic expansion of n!. J. Integer Sequences **13**(6), Article 10.6.6 (2010)
21. Nymann, J.E: On the probability that k positive integers are relatively prime. J. Number Theory **4**, 469–473 (1972)
22. Nymann, J.E: On the probability that k positive integers are relatively prime II. J. Number Theory **7**, 406–412 (1975)
23. Offner, C.D.: Some early analytic number theory. Accessed May 30, 2024. http://www.cs.umb.edu/~offner/files/an_num_th.pdf
24. Ramanujan, S.: The Lost Notebook and Other Unpublished Papers, Raghavan, S., Rangachari, S.S. (Eds.). Springer, New York (1988)
25. Riemann, B.: Über die Anzahl der Primzahlen unter einer gegebenen Grösse. Monatsberichte der Berliner Akademie, 671–680 (1859)
26. Smith, W.D.: The gamma function revisited (2014). https://schule.bayernport.com/gamma/gamma05.pdf
27. Spivak, M.: Topics in Calculus, 3rd edn. Publish or Perish, Houston (1994)
28. Yaglom, A.M., Yaglom, I.M.: Challenging Mathematical Problems with Elementary Solutions I, Translated by James Mc Cawley. Dover Publications, New York (1967)

Index

A
Abbott and Costello, 29
abelian group, **57**
absolutely convergent series, **119**
according to Hoyle, 71
ace on a die, **85**
Achilles, 12
Achilles heel, 13
acid
   deoxyribonucleic, 95
   nucleic, 95
   ribonucleic, 95
à Deux Tableux, 38
Adkins, Brian Lindsay, x
Alexandria, 151
algebra
   σ, **5**, **99**
alternating harmonic series, **144**
alternating sequence, **120**
amalgamation paradox, **46**
American Mahjong, 77
American roulette wheel, **18**
American Statistician, 103
ancient Romans, 88
André, Désiré, 41
André's reflection method, 41
Antiochus IV epiphanes, 86
ANZAC Day, **88**
A priori estimate, 106
Arccos, 122
arcsin, 122
arctan, 122
Ars Conjectandi, 108
Ask Marilyn, 104

associativity, **56**
asymptomatic, 98
asymptotic, **119**
Attack, Birthday, 89
Australian National University (ANU),
   x

B
Baccarat, 38
bacteria, 95
Ballot Box Problem, **40**
Ballot Box Theorem, 40
Barnes, Ernest William, 161
Bartlett, Maurice Stevenson, 45
Bartolini, Loretta, xvii
Bau cua ca cop, 29
Bayes' Theorem, **101**
Bayesian, 12
   inference, 105
   probability, 105
Beijerinck, Martinus, 95
bell-shaped, **124**
Benter, Bill, 35
Bernoulli brothers, 118
Bernoulli, Daniel, 126
Bernoulli, Jacob, 108
Bernoulli, Johann, 118
Bertrand, Joseph Louis
   François, 41
Bertrand's Ballot Box Theorem, 41
Bertrand's Box Paradox, **102**
Bertrand's box problem, **102**, 103
Bertrand's postulate, 102

bet
  Voisins du Zero, **49**
  Zero, **49**
bias
  survivor, xiv
Bible, 42
Bienaymé formula, **14**
Bienaymé, Irénée-Jules, 14
Binomial Distribution, 109
bird cage, **29**
Birthday attack, 89
Birthday paradox, 89
Birthday problem, 89
blackjack, **30**
Bletchley Park, 45
Bond, James, 27, 29
Bones
  Pussers, 77
bookie, **35**
bookmaker, **35**
Books, **75**
Borel, Félix Édouard Justin Émile, 7
Borel set, **7**
Borel σ-algebra, 7
botanist, 95
Boxer, **89**
Boxman, **28**
Bradbury, Ray, 43
Bramley-Moore, Leslie, 45
Breugel, Pieter the Elder, 87
Bridge
  contract, **71, 72**
Bridge whist, **71**
Burnside, William, 138
Bust, **31**
butterfly effect, 43

C
$\binom{n}{k}$, 66
χ^2-Distribution, 129
Cairo, 151
calculus, 114
caller, 82
Canasta, 75
Cantonese Mahjong, 77
Capell, Kenneth, x
Cardano, Gerolamo, xiv, 8
Carroll, Lewis, 87, 108
cartomancy, 16
Cavendish Road State High School, x
cell, 95
Chamberlain filter, 94
Chamberlain-Pasteur filter, 94
Chamberland, Charles, 94

Chamberland-Pasteur filte, 95
Chanukah, 86
chaos theory, 42–44
Chard in, Jean Baptiste-Siméon, 87
Chaucer, Geoffrey, 26
Chebyshev, Pafnuty Lvovich, 102
Chemin de Fer, 38
chi^2 test, 129
choose, 66
  in R language, **68**
chosen at radom, **157**
Chow, **78**
chromosome, 95
chuck-a-luck, **29**
clock
  pendulum, 138
coin
  fair, **9**
Collier's magazine, 43
combination, **66**
comparison test for infinite integrals, **121**
Compleat Gamester, 26
complement, **2**
composite number, 150
composition, **56**
conditional probability, **100**
confounding variable, 46
Confucius, 76
Connery, Sean, 27
Conquian, 75
Contract bridge, **71, 72**
converge, **116, 121**
convergent
  absolutely, **119**
  series, **119**
convicts, 88
coprime, **150**
Cos, 122
Cosh, 122
Costly Colours, 59
Cotton, Charles, 26
counter-intuitive, 104
couples
  strongly carefree, **159**
craniology, 45
Craps, 26
craps, 27, 28
Cribbage, 59, 60
Crick, Francis Harry Compton, 96
cricket, 88
Cross and pile, 88
Crown and Anchor, **29**
crystallographer, 95
  X-ray, 95

Index 167

Csc, 122
culture, 97
cycle, **56**
cystic fibrosis, **19**

D
data
　interpretation of, xiv
Davis, Clive, x
de Moivre, Abraham, 114
de Montmort, Pierre Remond, 161
De Morgan, Augustus, 3
De Morgan's Laws, **3**
dead heat
　triple, 35
dead man's hand, **83**
Deck of cards, **15**
Dedekind cut, 102
Dedekind, Richard, 102
deoxyribonucleic acid, 95
derangement, **161**
Diamonds are Forever, 27
dice, **10**
Dickens, Charles John Huffam, 87
die, **10**
　fair, **10**
differentiable, **132**
　infinitely, **132**
disease
　Gaucher's, **47**
　monogenic, **19**
　recessive, **19**
disjoint
　pairwise, **8**
Distribution
　binomial, 109
　gamma, 109, 128
　Gaussian, **124**
　normal, 109
　Poisson, 109
　standard normal, **124**
distribution
　χ^2, 129
　Erlang, 128
　exponential, 129
　hypergeometric, 81, 109
　probability, **8**, **99**
　Yule, 45
　Yule-Simon, 45
divergent, **116**, 119
diverges, **116**
divisor, **150**
　greatest common, **150**
DNA, 95

Dodgson, Charles Lutwidge, 87
dominate, **121**
don't pass, 28
Double, **31**
Dreidel, **86**
Durham, Joseph, 151

E
Egypt, 151
Egyptologist, 151
Elements, 151
　Euclid's, 151
eloped, 60
engineering
　traffic, 128
ensemble forecasting, 44
enzyme, 97
Erdös, Paul, 133
Erlang distribution, 128
Erlang, Agnet Krarup, 128
$E \setminus S$, **2**
Euchre, 71, 74
Euclid, 102, 151
Euclid's Elements, 151
Euclidean Geometry, 151
Euler, Leonhard, 123, 144
Euler-Poisson integral, **123**
European roulette wheel
　European, **18**
event space, **8**, **99**
E(X), 40
exclusive
　mutually, **8**
expectation, **10**
expected value, **10**, 40
exponential distribution, 129
exponentiation in R language, **68**
extraneous variable, 46
Eyes, **78**

F
fair coin, **9**
fair die, **10**
false negative, 98
false positive, 98
Fibonacci, 23
filter
　Chamberland, 94
　Chanmberland-Pasteur, 94, 95
Finucan, Henry, xi
first Scottish war, 60
First World War, 88
Fisher, Sir Ronald Aylmer, 12
Five hundred, 71

fixed point, 161
Flinders University, x
flu, 95
flush, **69**
  royal, **69**
  straight, **70**
forecasting
  ensemble, **44**
formula
  Bienaymé, 14
fortune-telling, 16
Foyle's War, 29
France, 60
Franklin, Rosalind Elsie, 95, 96
Freehold Raceway, 35
French roulette wheel, **18**
full house, **69**
function
  Gamma, **126**
  injective, **4**
  inverse, **4**
  one-to-one, **4**
  Riemann zeta, **144**
  surjective, **4**
Fundamental theorem of algebra, 144

G
Gallipoli Campaign, 88
gamblers' fallacy, 14
Gamma distribution, 109, 128
Gamma function, **126**
Gani, Joseph, xi
Gardner, Martin, 108
Gaucher, Claud Ernest, 47
Gaucher's disease, **47**
Gauss, Johann Carl Friedrich, 123
Gaussian distribution, **124**
Gaussian integral, **123**
gelt, 86
Genesis, xvii, 42
Genetics, 12
Geometry, 151
  Euclidean, 151
geometry, 151
Gin Rummy, 75, 76
Goldbach, Christian, 126
gold standard, 97
Gosper, Ralph William Jr., 138
grand hazard, 29
greatest common divisor, **150**
Greek alphabet, 4
Grenfell, Bernard Pyne, 151
grosser Schlag, 38
group, 57

abelian, **57**
permutation, **57**

H
Hall, Monty, 103
Han dynasty, 42
Harbin Mahjong, 77
hard, **31**
harmonic series, **120**
Hart, Vincent Gerald Michael, x
Hasofer, Abraham Michael, xi
Hazard, **26**
Heads, **89**
Hebrew Bible, 86
Helenistic, 86
Helmert, Friedrich Robert, 129
Hermite, Charles, 41
Hilbert, David, 1
Hillel sandwich, 59
Hirschhorn, Michael, 140
Hit, **31**
HIV, 99
Hong Kong Mahjong, 77
Hoo Hey How, **29**
horse racing, 35
Housey-House, 82
Howard, Lepo Esmond, x
Hoyle, Edmond, 71
human immunodeficiency, 99
Hunt, Arthur Surridge, 151
Hun-Tun, 42
Huygens, Cristiaan, 138
hypergeometric distribution, 81, 109

I
identity permutation, **56**
Il Gioco del Lotto d'Italia, 82
image
  inverse, **4**
improper integrals, 114, **120**
independent random variables, **12**
inference
  Bayesian, 105
infinite integral, 114, **120**
Infinitely differentiable, **132**
infinite product, **152**
infinite series, 114, **119**
influenza, 95
injective function, **4**
Inquisition, 60
insurance, **31**
integer part, **153**
integral
  Euler-Poisson, 123

Gaussian, **123**
improper, **120**
infinite, **120**
Riemann, 146
integration by parts, **127**
interpretation of data, xiv
intuition, 24
inverse function, **4**
inverse image, **4**
Israel Mahjong, 77
Ivanovsky, Dmitri Iosifovich, 95

J
Jacobs, Konrad, xii
Japanese Mahjong, 77
Jerusalem, 86
Jewish, 86
Joker, 74
Jones, Alan Stuart, x
Jones, Graham, x
Journal of Royal Statistical Society, 45

K
King Charles 1, 60
Klah Klok, **29**
kleiner Schlag, 38
Klondike Solitaire, **47**
Kluvanek, Igor, x
Knave, 59
Kolmogorov, Andrej Nikolajewitsch, 1, 100
Kong, **78**

L
La Condamine, Charles Marie, 81
Langur Burja, 29
La Partage, 18
Laplace, Pierre-Simon, marquis de, 10, 123, 139
Latham, Robert, 84
Le Lotto, 82
Lee, Alice, 45
Legendre, Adrien-Marie, 126
Leibniz, Gottfried Wilhelm, 85, 108
Leibniz-Newton calculus controversy, 108
Leonardo of Pisa, 23
L'Hôpital's rule, **117**, 118
lim, **116**
limit comparison test for infinite integrals, **125**
limit of a sequence, **116**
Lipton, Stephen, x
ln, 122
Lorenz, Edward Norton, 44
lotteries, 80
lotto, **80**, 84

Louisiana, 74

M
Macao, 37
Maccabean, 86
Macdonald, Ian D., x
Maclaurin, Colin, 131
Maclaurin series, **132**
Maclisp, 138
Macsyma, 138
magazine
    Collier's, 43
Magie, Elizabeth, 14
Mahjong, 76, 77, 80, 89
    American, 77
    Cantonese, 77
    Harbin, 77
    Hong Kong, 77
    Israel, 77
    Japanese, 77
    Three Player, 77
    Vietnamese, 77
Mahjong Solitaire, 77
Mahjong, Tianjin, 77
mapping
    one-to-one, 56
marked playing cards, 60
Markov, Andrey Andreyevich, 109
Markov chain, 109
Markov process, 109
Martingale betting system, **23**
Martyn, Marguerite, 71
matching problem, **161**
mean, 124
measure
    probability measure, 8, 99
Melbourne Cup, 35
Melds, 75, **78**
Mengoli, Pietro, 144
Messenger of Mathematics, 40
Met Office, 44
microbiologist, 95
micrometre, 94
micron, **94**
Miescher, Johannes Friedrich, 96
misogyny, 104
Mississippi River, 74
misuse of statistics, **xiii**
monogenic disease, **19**
Monopoly™, 13
Monte Carlo analysis, 44
Monty Hall Problem, 103, 107, 108
Moore, Roger, 29
Mullis, Kary Banks, 97

multiplication in R language, **68**
mutually exclusive, **8**
mutually prime, **150**

N
natural, **28**, 39
navia aut caput, 88
Navy
  Royal Australian, 77
Nemes, Gergő, 139
Neumann, Bernhard H., x
Neumann, Hanna, x
New Orleans, 74
Newton, Isaac, 85
New York Times, 97
Nichols, Des, x
Nobel prize, 45
Noddy, 59
Normal distribution, 109
  standard, **124**
nucleic acid, 95
number
  composite, 150
  Fibonacci, 23
  prime, **48**, **150**
  supplementary, **84**

O
Oberwolfach, xii
occult, 16
Odds, **35**, **89**
one-to-one function, **4**
one-to-one mapping, 56
onto, **4**, 56
Oresme, Nicole, 120
Oxyrhynchus, 151

P
2^Ω, **5**
$P(A)$, **8**, **99**
P_k^n, 66
Pólya's Urn, **109**
Pólya, George, 109
Pair, **78**
pairwise disjoint, **8**
Papyrologist, 151
Parade magazine, 103
paradox
  amalgamation, 46
  reversal, 46
  Simpson's, 46
parameter
  scale, 128
  shape, 128

Paris Academy of Sciences, 102
Pascal, Blaise, 161
Pascal's triangle, 161
pass, 28
Pasteur, Louis, 94
Patience, **47**
PCR test, 97
Pearson, Karl, 45, 129
penal colonies, 88
pendulum clock, 138
Pepys, Samuel, 84
perfect hand, 64
permutation, **56**
  group, **57**
  identity, **56**
Phar Lap, 35
Poe, Edgar Allen, 87
Poincarè
  Jules Henri, 42
point, 28
  fixed, 161
Poisson, Siméon Denis Poisson, 123
Poisson distribution, 109
Poker, 65, 66
poker face, 70
polymer, 97
Polymerase chain reaction, 97
Pong, **78**
pontoon, **30**
population standard deviation, **12**
population variance, **12**
posterior estimate, 106
Postulate
  Bertrand's, 102
powerball, **81**, 84
power set, **5**
prime
  mutually, **150**
  relatively, **150**
prime number, **48**, **150**
Principia Mathematica, 85
Principle of Inclusion and Exclusion, **157**
probability, **8**, **99**
  Bayesian, 105
  conditional, **100**
  of B given A, **100**
  distribution, **8**, **99**
  measure, **8**, **99**
  space, **8**, **99**
Problem
  Monty Hall, 103
  birthday, 89
  urns, 109
Proof by contradiction, 152

Index

Punto Banco, 38
Pussers Bones, 77

Q
Qing dynasty, 76
quadratfrei, **158**
queueing theory, 128
Quinion, Michael, 27

R
Ramanujan, Srinavasan, 138, 161
random variables, **10**
  independent, **12**
  uncorrelated, **14**
recessive disease, **19**
Relatively prime, **150**
reversal paradox, 46
ribonucleic acid, 95
Rieman, Georg Friedrich
    Bernhard, 146
Riemann integral, 146
Riemann zeta function, **144**
RNA, 95
roulette, **17**
roulette wheel
  American, **18**
  French, 18
Royal Australian Navy, 77
royal flush, **69**
rule
  L'Hôpital's, **117**
Rummikub, 77
Rummy, **75**
  Gin, 75
Runs, **75**

S
S^c, **2**
σ-algebra, **5**, **99**
  Borel, 7
  trivial, **5**
σ, 11
Sachs, Bernard, 20
Salvin, Steve, 103
sample space, **8**, **99**
sample standard deviation, **12**
sample variance, 11, **12**
sample with replacement, **66**
sample without replacement, **65**
Sanskrit, 23
Saturn, 138
scale parameter, 128
Scientific American, 108
Sec, 122

Second Temple, 86
selected, **65**
  with replacement, **65**
  without replacement, **65**
selection
  with replacement, 108
  without replacement, 108
Seleucids, 86
sensitivity, **98**
sequence, **116**
  alternating, **120**
  asymptotic, **119**
  convergent, **116**
  divergent, **116**
series
  absolutely convergent, **119**
  convergent, **119**
  harmonic, **120**
  infinite, **119**
  Maclaurin, **132**
  Stirling, **139**
  Taylor, **132**
set, **75**
  Borel, 7
  power, **5**
sexism, 104
Shakespeare, 26
Shanghai Solitaire, 77
shape parameter, 128
shooter, 26
sic bo, 29
sickle-cell anaemia, **19**
Simon, Herbert Alexander, 45
Simpson, Edward, H., 45
Simpson's paradox, **46**
sin, 122
sinh, 122
sixes and sevens, 27
smooth function, **132**
soft, **31**
Solitaire, **47**
  Klondike, **47**
  Mahjong, 77
  Shanghai, 77
space
  event, **8**, **99**
  probability, 8, 99
  sample, **8**, **99**
Spain, 60
specificity, **98**
Spinner, **89**
Split, **31**
Springer Nature, xvii
St Louis, 71

St Louis Post Dispatch, 71
Stand, 31
standard deviation, **11**, 124
  population, **12**
  sample, **12**
Standard normal distribution, **124**
standard, gold, 97
statistics
  misuse of, **xiii**
Steinsaltz, Rabbi Adin, 6
Stickman, **28**
Stirling, James, 114
Stirling's approximation formula, **114**
Stirling series, **139**
straight, **70**
  flush, **70**
Street, Anne Penfold, x
strongly carefree couples, **159**
Suckling, Sir John, 60
suit, **15**
sum of infinite series, **119**
supplementary numbers, **84**
surjective function, **4**
survival bias, **xiv**
sweat rag, 29
system
  Martingale, **23**

T
tan, 122
tanh, 122
Tails, **89**
Taoism, 42
tarot cards, **16**
Tay, Warren, 20
Tay-Sachs disease, **20**
Taylor, Brook, 127, 131
Taylor series, 115, **132**
Technical University of Darmstadt, xi
teetotum, 87
test
  χ^2, 129
  Polymerase Chain Reaction, 97
The Landlord's Game, 14
Theorem
  Bayes', **101**
  Bertram's Ballot Box, 41
theory
  chaos, 42
  queueing, 128
Three Player Mahjong, 77
Tianjin Mahjong, 77
Titus, 138

tobacco mosaic virus, 95
Tombola, 82
topology, **7**
Torah, 86
traffic engineering, 128
triangle, Pascal's, 161
trick, **71**
trick-taking game, **71**
trigonometry, 122
Trinity College, Dublin, 45
Triple dead heat, 35
trivial σ-algebra, **5**
Trump, **71**
Tulane University, xi
twenty-one, **30**
Two-Up, **88**, 89

U
uncertainty, 105
uncorrelated random variables, **14**
United Kingdom, 44
  Oxford University, 151
  University of Cambridge, 45
  University of Queensland, x
  University of Vienna, xiv
  University, Oxford, 151
Urn
  Pólya's, **109**
  problems, 109

V
Výborný, Rudolf, x
Var(X), 11
variable
  confounding, 46
  extraneous, 46
  random, **10**
variance, **11**
  population, **12**
  sample, 11, **12**
verkauft, 38
Vietnamese Mahjong, 77
Villarino, Mark B., 140
vingt-un, **30**
virology, 95
virus, **94**, 95
  tobacco mosaic, 95
Voisins du Zero bet, **49**
Voltaire, 81
von Mises, Richard Edler, 89
vos Savant, Marilyn, 103

W
Wald, Abraham, xiv
Watson, James Dewey, 96

Watts, Anthony McLean, x
Whist, **71, 72**
  Bridge, **71**
Whitworth, William Allen, 40
WHO, 19
Wilde, Oscar, 57
Williams, Sheila, x
Windschitl, Robert H., 139
Women's Whist Club Congress, 71
World Health Organization, 19

X
X-ray crystallographer, 95

Y
Yarborough, **72**
  2nd Earl of, 72
Yiddish, 86
yo, **29**
Yule distribution, 45
Yule, George Udny, 45
Yule-Simon distribution, 45
Yule-Simpson effect, **46**

Z
Zero bet, **49**
zeta function
  Riemann, **144**

SPRINGER NATURE

GPSR Compliance

The European Union's (EU) General Product Safety Regulation (GPSR) is a set of rules that requires consumer products to be safe and our obligations to ensure this.

If you have any concerns about our products, you can contact us on ProductSafety@springernature.com

In case Publisher is established outside the EU, the EU authorized representative is:

Springer Nature Customer Service Center GmbH
Europaplatz 3
69115 Heidelberg, Germany

The manufacturer's authorised representative in the EU is Springer Nature Customer Service Centre GmbH, Europaplatz 3, 69115 Heidelberg, Germany. If you have any concerns regarding our products, please contact ProductSafety@springernature.com

Printed and bound by CPI Group (UK) Ltd, Croydon, CR0 4YY
26/03/2026
02078974-0002